THE ROLE OF AGING IN ATHEROSCLEROSIS

THE ROLE OF AGING IN ATHEROSCLEROSIS

The sequestration hypothesis

by

Richard E. Tracy MD PhD

Department of Pathology,
Louisiana State University Health Science Center,
New Orleans, U.S.A.

KLUWER ACADEMIC PUBLISHERS
DORDRECHT / BOSTON / LONDON

A C.I.P. Catalogue record for this book is available from the Library of Congress.

ISBN 978-90-481-6265-9

Published by Kluwer Academic Publishers,
P.O. Box 17, 3300 AA Dordrecht, The Netherlands.

Sold and distributed in North, Central and South America
by Kluwer Academic Publishers,
101 Philip Drive, Norwell, MA 02061, U.S.A.

In all other countries, sold and distributed
by Kluwer Academic Publishers,
P.O. Box 322, 3300 AH Dordrecht, The Netherlands.

Printed on acid-free paper

CONTENTS

PREFACE

The cover of this book summarizes the central features of the sequestration hypothesis: Commonplace appearances seen in human coronary artery, fat stained in paraffin sections by a new technique explained in Chapter Eleven, are arranged to suggest pathways of evolution toward atheroma. The hypothesis formulated and defended in the pages of this book is this: Fibroplasia progresses upward in column "a" from "1a" to "3a" as a characteristic feature of aging. This starts sooner and progresses faster in men than in women. Numbers of SMC's remain essentially constant so that fibroplasia per SMC steadily increases. The rise upward conveys an increasing propensity to sequester atherogenic lipids, causing transition rightward into column "b". Sequestered extracellular lipid then attracts fatty streak elements, especially foam cells and lymphocytes, to propel the arterial site rightward into column "c". Frame "1c" corresponds to the AHA Lesions Committee classification type IIb, the progression resistant fatty streak arising directly without prior lipid sequestration; this can progress to atheroma, but slowly after much delay, although extreme provocation can accelerate the process. Such progression is rightward toward atheroma with thin cap, not upward toward fibroplastic thickening. Frame "2c" corresponds to the AHA classification, type IIa, progression prone fatty streaks. These readily evolve into atheroma, again by horizontal progression. Frame "3a" corresponds to the AHA classification "adaptive thickening", which becomes type III lesions in Frame "3c" after sequestering lipid and then acquiring fatty streak elements; the deeply situated extracellular lipid is conspicuous here. These are "intermediate lesions" in the sense of standing between Frame "3a" and atheroma (not illustrated here). They are not intermediate in the sense of standing between frames "1a" and "3c", a form of evolution that is thought to happen only in the fibrous cap of the atheroma, after the intrusion of a necrotic core into the prepared site. The risk factors age and maleness are thought to act upon arterial intima, to prepare for atheroma, by propelling fibroplasia upward in column "a"; geography is also envisioned in this way, as for instance the contrast between New Orleans and Mexico City. The great statistical power of age, maleness, and geography as risk factors for atherosclerosis suggest that the upward progression in column "a" is the rate limiting step in atherogenesis under usual circumstances in most persons. Arrows mark intima-media boundaries, these arrows are all of length 20 μm.

PREFACE

INTRODUCTION

HOW TO MEASURE AGE

In human experience, the dominant risk factor for atherosclerosis and its complications is age. Yet age rarely enters the usual experimental models of atherosclerosis in laboratory animals, and even then only as a peripheral concern. The present inquiry, therefore, concentrates solely upon human experience, since animal observations have so little immediate relevance.

In most studies of aging, the age of an animal is measured not in units of days or years, but rather in fractions of the usual life span of the species [15]. A dog year is said to be seven man years, and this would point to one rat year as 28 man years, and one mouse year as 56 man years. This way of measuring is not of great relevance to the aging of arteries. Rather, we have reason to suspect that the 3-year old rat resembles the 3-year old human, the 8 year old cow resembles the 8 year old human, the 20-year old horse resembles the 20-year old human, and so on [3,59]. This idea is based upon small amounts of data and some limited observations in descriptive pathology. If it should prove correct, however, then experimental approaches to explaining the "age" risk factor will be tedious, slow, and expensive.

Investigators seldom pose the question of what does aging do to the arterial tissue in preparation for later intrusion by atheroma. Perhaps the reason for this omission is that plausible hypotheses are wanting, leaving a vast gap in our understanding. Yet that gap is now beginning to fill, because of the observations to be examined in this treatise.

An especially crucial feature of arterial aging is the time-ordered incremental thickening of the intimal layer, beginning at or before age 20 years around the cessation of pubertal growth. Concerning this condition, Klotz in 1911 [26] tells us, "In 1885,Thoma set forth his views concerning the fundamental principles underlying arteriosclerosis. From the first Thoma's attention was concentrated on the intimal thickening . . . and with the degenerative processes that develop in these intimal thickenings." This theory later evolved into generalized arteriosclerosis [1;Allbutt 1911], and has persisted under such names as inelastic fibrous tissue [17; Duguid 1926], intimal thickening [16; Duff 1935], arterial injury [14; Dock 1946], diffuse intimal thickenings (DITs) [80; Wilens 1951], eccentric thickenings [48; Stary 1992], adaptive intimal thickenings [10,11; Committee on Lesions of the AHA 1992], and pathological thickenings [79; Virmani et. al., 2000]. In this discussion the term "fibroplastic arteriosclerosis" is preferred, bearing in mind that it carries an implication of generalized affliction of all arteries.

The observations reviewed in future chapters generate two startling conclusions: (1) The influence of age on the evolution of fibroplastic arteriosclerosis can be summarized in an amazingly simple formula, $F = \beta A^2(1 - \alpha A)$, where F is a quantitative measure of fibroplastic arteriosclerosis, and A is age in years. The coefficients β and α are empirical parameters that serve to transform the effects of time, A, into their morphological consequences, F. The parameters vary between groups of subjects such as between men and women and between nations, so that "aging" is not exactly identical in everyone nor in every group. Aging of arteries cannot be measured in fractions of life span, but it also cannot be measured by the strict ticking of an exact atomic clock. (2) The intrusion of atheroma is governed by another amazingly simple formula, $W = bF - a$, where $W > 0$ marks a high probability of atheroma and $W < 0$ marks a low probability, while a and b are empirical parameters. The parameters a and b can be influenced by fatty deposits (i.e. "fatty streaks"), arterial size, and some other factors. Current evidence, however, finds remarkable constancy under most circumstances for the *ratio* of the parameters a/b, which determines F when $W = 0$. This ratio fails to vary among age groups, between demographic groups, and in other comparisons. This constancy implies that the effect of age upon intrusion of atheroma is governed solely by its actions upon F in all kinds of human subjects.

The way that age acts to prepare the arterial intima for intrusion of atheroma is now known in broad outline, and the search can now begin for the biological mechanisms to explain this phenomenon. By knowing what agencies affect the magnitudes of which parameters in the system, we can devise testable hypotheses. For instance, data reviewed later indicate a possible strong effect on the parameter β by arteriolar hyalinization in the renal cortex. Subjects with severe hyalinization appear to grow old more quickly than usual in the particular tissues of the arteries. This could be the starting point for fruitful inquiries.

Chapter One

Introducing the Sequestration Hypothesis

Abstract. Aging of the coronary intima sometimes induces SMC's to fabricate excessive collagenous matrix materials around themselves. Could this fabrication be the usual rate limiting step governing the evolution toward atheroma?

1.1. Photo Illustrations of Hypothesis

Figure 1-1 provides examples of features frequently seen in H&E stained paraffin sections of human coronary arteries. These are arranged to illustrate a proposed pathogenetic scheme [53-56], as summarized here in five stages: (1) During adolescence the coronary artery is thought to construct a normal intima which further matures until about age 25 years (Figure 1-1A). (2) During its ongoing maturation, this intima begins to acquire small regions of fibroplastic thickening, sometimes called adaptive thickening [10,11,48], which display excessive production of collagenous matrix by a nearly stable population of smooth muscle cells (SMC's) (Figure 1-1B).Such fibroplasia begins at well defined anatomic sites [11,49] and spreads from there throughout life, often coming to occupy the entire coronary artery [60,61]. (3) Scattered deposits of lipid arise, as recognized in H&E-stained paraffin sections by the

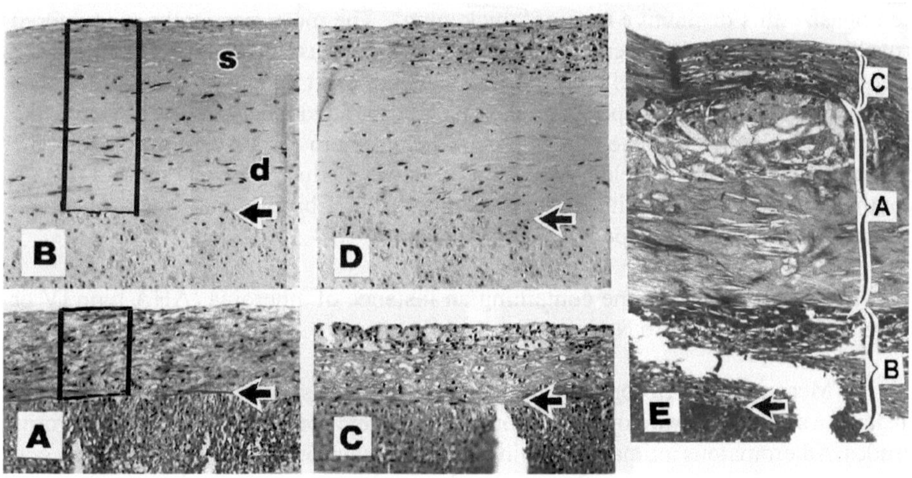

Fig 1-1. Examples of prevalent appearances in coronary arteries are arranged to suggest a pathogenetic scheme. Arrows mark intima-media boundaries. Rectangular areas of 100μm width in A and B are used to enumerate SMC's. Lettering in Frame E identifies the atheronecrotic lipid core, A, and its fibrous cap, C, and fibrous base, B. H&E.

presence of foam cell infiltrates accompanied by lymphocytes and other fatty streak elements [48], these can affect regions of previously normal intima (Figure 1-1C) or variably far advanced fibroplasia ("adaptive thickening", Figure 1-1D). (4) When fatty streak elements collocate with fibroplasia, this produces a progression prone condition that often quickly evolves into outright atheroma (Figure 1-1E). When deposited into normal intima, fatty streak elements can initiate an evolution to atheroma, but only with extreme provocation. (5) Atheroma, i.e. the necrotic core which is the chief hallmark of the "vulnerable plaque", can sometimes precipitate ischemic injury and death.

Step 2 may be the usual rate limiting step in the proposed scheme of atherogenesis. Advanced degrees of fibroplastic intimal thickening, Step 2, are thought to act as "lipid traps" [45,78,80,81] to catch and hold atherogenic lipids, and this concept will be elaborated in later chapters. The intimal fibroplasia at Step 2 can be laid down by excessively numerous SMC's or through excessive production of fibrous matrix materials by each of the SMC's. Previous reports [54] described a dominant influence by the second kind of fibroplasia, excessive matrix per SMC, in the promotion of atheroma and cardiac death.

1.2. Atheromatous and fibroplastic intimal thickenings

Figure 1-1E illustrates *atheromatous* thickening and Figure 1-1B illustrates *fibroplastic* intimal thickening of coronary arteries. These two distinct pathological entities call for careful separation. To help achieve this objective, the term "fibroplasia" is introduced to emphasize this distinctive type of "thickening". The more frequently encountered terms "hyperplasia" or "hypertrophy" are avoided, because those terms can hold the deceptive implication that cells proliferate, which doesn't happen in this setting. The recently popular term "adaptive thickening" is avoided because it can have the unwanted implication that the process is confined only to certain anatomic sites. The collocating of fatty streak elements with fibroplasia, illustrated in Figure 1-1D, is hypothesized to be a transition stage leading from fibroplasia to atheroma, and corresponds to the "AHA type III" lesion [11].

A YesA artery is one containing an instance of atheroma (AHA type IV or greater [11]); all others are NoA arteries.

Measuring the fibroplastic intimal thickness in a specimen of coronary artery is usually a straight forward operation, but it encounters complications when atheroma intrudes. Atheromatous intimal thickenings often alter surrounding conditions in several ways, including calcification, inflammation, hemorrhage, vascularization, erosion, and wound healing responses related to SMC proliferation. For this reason, the atheromatous portions of artery are omitted when measuring fibroplasia, and the measured sample must therefore be treated statistically as a "censored" data set, as examined in Chapter Three and later.

1.3. Step 2, emergence of "fibroplastic arteriosclerosis"

When the normal intima acquires fibroplasia, the numbers of SMC's are found to remain nearly stable [53-56]. This is illustrated in Figure 1-1 by comparing Frames A and B. The demarcated rectangular region of coronary intima in Figure 1-1A contains 28 nuclei of SMC's within the 140 μm of intimal thickness, giving 140/28 = 5.0 μm/SMC, which exemplifies the average "normal". The rectangular region in Figure 1-1B contains 27 nuclei of SMC's within the 310 μm of thickness, yielding 310/27 = 11.5 μm/SMC, which exemplifies the condition "fibroplastic arteriosclerosis".

The proposal here is to diagnose the disease entity "fibroplastic arteriosclerosis" in tissue sections as a magnitude in excess of "normal". Later chapters confront the challenge of determining what is "normal", and the outcome is summarized here. The value 3.3 μm/SMC emerged from regression equations as the magnitude of normal fibroplastic intimal thickness at age 20 years. The values of this quantity obtained in subjects of ages 15-24 years ranged from 2.6 to 8.5 with an average of 5.0 μm/SMC, and this offers a second estimate of the normal value. The following discussion concerns how to estimate the upper limit of normal.

1.4. Components of coronary intimal fibroplasia

The fibroplastic thickness of coronary intima, F, is a product of two independent factors, C_T = total number of smooth muscle cells (SMC's) in a rectangular area such as Figure 1-1A or B, and F_C = fibroplasia per SMC as just discussed. $F = C_T*F_C$. If the total thickness of fibroplastic intima is what acts as a lipid trap, then the two components might each contribute to the observed effect, but the contributions might be of unequal importance. This matter is examined next.

1.5 Influences of two fibroplasia components upon foam cell infiltration in coronary intima

P_F denotes the percentage of a coronary artery affected by foam cell infiltrates, such as those in Figures 1-1C&D. Figure 1-2 reports findings in coronary arteries of 194 men aged 35-79 years [54]. High P_F and Low P_F groups were constructed as follows: Within each 10 year age range, High P_F subjects are defined as those above the median and Low P_F subjects are below the median in foam cell infiltration. The graph shows High P_F subjects tending to have the greatest fibroplasia per SMC, F_C, (circles tend to fall rightward of the diagonal in the graph) but also to have the greatest numbers of SMC's, C_T (upward from the diagonal). The importance of F_C is clearly greater than that of C_T, because the diagonal line is nearly vertical. Indeed, the contribution of C_T is ambiguous, because of weak statistical significance and because of certain concerns for sampling bias to be discussed in Chapter 4; whether its contribution significantly differs

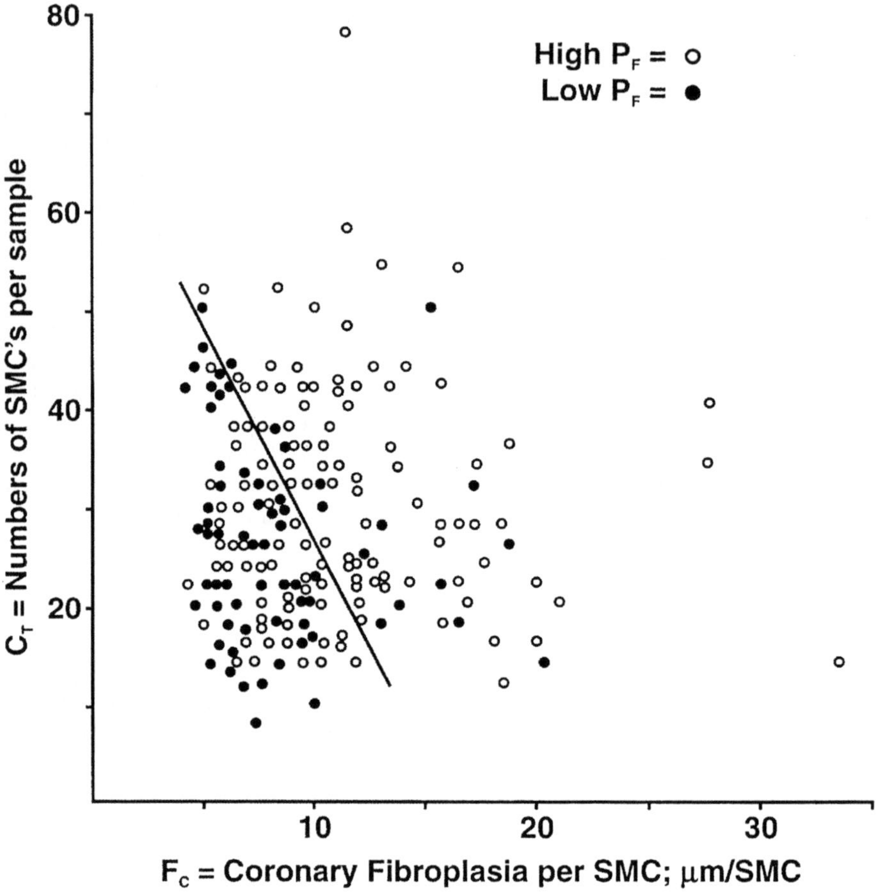

Fig 1-2. Each symbol represents the means of P_F and F_C observed in an autopsy. The nearly vertical solid line represents Equation 1-1, the discriminant function that optimally separates circles from dots. Men only, aged 35-79 years with all causes of death. P_F measures foam cell infiltrates in the nonnecrotic portion of each specimen; High and Low are respectively above and below the age-specific medians.

from zero calls for further data to resolve this important issue. Figure 1-2 offers evidence that intimal fibroplasia acts as a lipid trap, because foam cell infiltrates are recognized operationally as the visible manifestation of lipid accumulations in the paraffin sections studied here. The purist might object that perhaps lipid propels the fibroplasia, but this suggestion cannot be reconciled with the fact that fibroplasia is a generalized property found throughout the artery, even at sites remote from foam cells and atheroma (this matter is taken up in Chapters Three & Four, and further explored

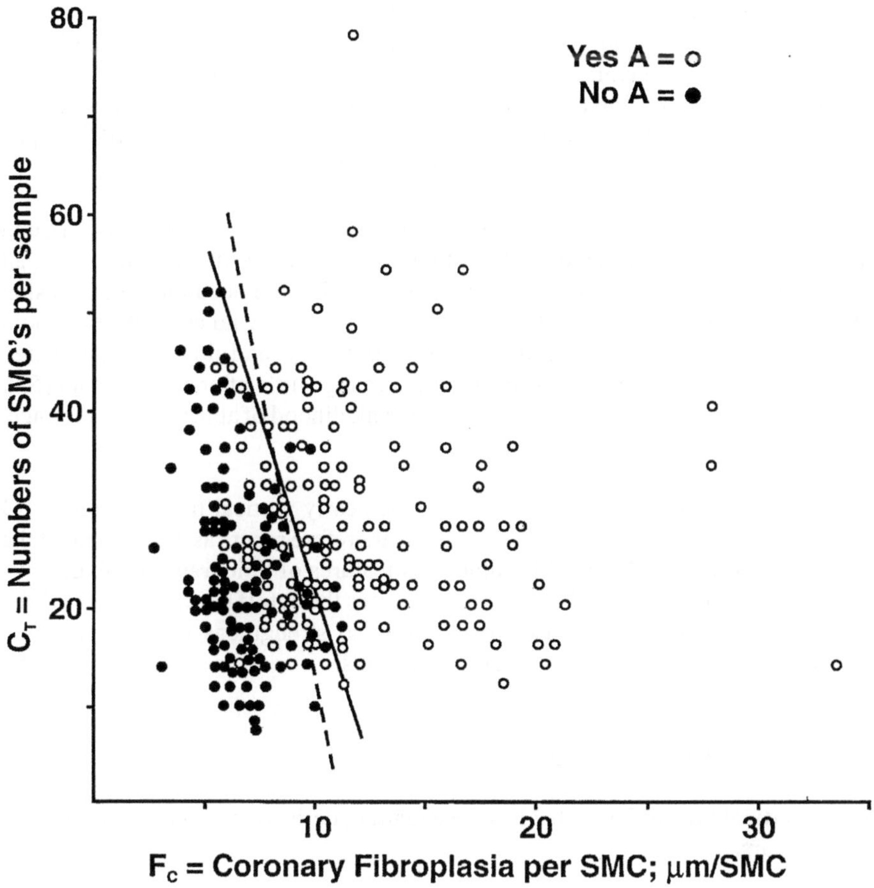

Fig 1-3. As in Figure1-2. Sloping lines represent versions of Equation 1-3 separately determined for men (solid) and women (dashed). Men and women aged 35-79 years with all causes of death.

in later chapters).

The nearly vertical line in Figure 1-2 centers around $F_C = 9.3$ µm/SMC, and this value is therefore an estimate of the upper limit for normal F_C. Above this value, the features of lipid trapping becomes likely, and this property is expected for fibroplastic arteriosclerosis [1].

1.6. Influences of two fibroplasia components upon atheroma in coronary intima

Figure 1-3 presents findings for men and women combined, because the sex groups do not differ significantly in this relationship. This similarity of the sex groups is illustrated by the nearness of the solid line for men to the dashed line for women. Arteries that contain instances of *atheromatous* intimal thickening, YesA in Figure 1-3, tend to be the ones that also have the greatest *fibroplastic* intimal thickenings generally throughout the specimen. The relationship of fibroplastic thickening to atheroma is dominated by fibroplasia per SMC, with little participation by excessive numbers of SMC's (open symbols tend to fall rightward from the nearly vertical lines in Figure 1-3).

Figure 1-3 offers a dynamic image of how atherogenesis may proceed: Young men and women begin near the left side of Figure 1-3, bearing little coronary intimal fibroplasia, and later move rightward as they grow older and accumulate increasing amounts of fibroplasia per SMC. The faster moving subjects who cross the threshold marked by the nearly vertical lines gain excessive fibroplasia, and along with this a high probability of acquiring atheroma. SMC numbers change little after age 35 years [53], and their numbers have minimal influence upon the likelihood of atheroma at every age, as demonstrated later.

As with the foam cells of Figure 1-2, the nearly vertical lines in Figure 1-3 again center around $F_C = 9.3$ µm/SMC. This value therefore recurs as an estimate of the upper limit for normal F_C. Above this value both features of lipid trapping, foam cells and atheroma, become likely. This outcome reinforces the cutoff point of 9.3 µm/SMC for diagnosing fibroplastic arteriosclerosis, and also strengthens the proposal that intimal fibroplasia acts as a trap for atherogenic lipids.

1.7. Use of discriminant function analysis

The nearly vertical line in Figure 1-2 has the equation, derived from a pool of 194 men aged 35-79 years,

$$W_P = 0.71F_C + 0.40C_T \quad (T^2 = 0.60 \quad \% = 67 \quad c=-2.59) \qquad (1\text{-}1)$$

This is the discriminant function that maximally distinguishes High P_F from Low P_F subjects when $W_P = 0$. Standardized coefficients are given so that variables are measured in units of their standard deviations. T^2 is the difference of mean W_P in the High P_F from Low P_F subjects measured in the squared standard deviation of W_P, % is the percent of cases correctly classified, and c is the constant to add when restoring raw coefficients through division by the standard deviations in Table 1-1.

$$W_P = 0.77F_C + 0.64C_T \quad (T^2 = 0.56 \quad \% = 65 \quad c=-2.83) \qquad (1\text{-}2)$$

Equation 1-2 presents the similar analysis for 81 women of ages 35-79 years. The comparable graph for the women differs little from that for the men, and is omitted for brevity; the equation summarizes the conclusions. Although the differences between

Equations 1-1 and 1-2 are small, they are of statistical significance, so that the sex groups ought to be handled separately.

Equation 1-3 is the discriminant function for optimally distinguishing circles from dots in Figure 1-3.

$$W_A = 2.01F_C + 0.67C_T \quad (T^2 = 2.69 \;\; \% = 82 \;\; c=-5.81) \tag{1-3}$$

Versions of equation 1-3 with $W_A = 0$, separately determined for men and women, are plotted in Figure 1-3, using a solid line for men and a dashed line for women. These are downward sloping, nearly identical to each other, and not significantly different. Separate equations for men and women are not significantly different, and a single equation can serve equally well for both groups. Symbols falling upward and rightward of the sloping lines are predicted to be YesA. Of the 127 YesA cases in the scatter plot for men, 94 have $W_A > 0$ (74 %) and are correctly predicted this way. Symbols falling downward and leftward from the solid line are predicted to be NoA, and the 62 with $W_A < 0$ among the 64 NoA men (97 %) are correctly predicted in this way. Among women, 63% of YesA and 93 % of NoA subjects are correctly predicted by equation 1-3. Emphasis is given here to the finding that **age is rejected from inclusion in equation 1-3** as statistically not significant in these subjects of 35-79 years of age. Although the YesA subjects are generally older than the NoA subjects (respective means of 52.4 vs 45.3 years), the difference in ages is statistically displaced by the variables included in equation 1-3. This finding raises the possibility that **age acts by way of the retained variables, which implies that we know what age does** to the tissues in preparation for the intrusion of atheroma. A similar conclusion is implied for the maleness risk factor; the maleness factor exaggerates intimal fibroplasia in men (especially the ominous F_C linked form to be elaborated in later chapters) in comparison with women, and no other action of "maleness" is statistically required to "explain" the excessive production of atheromas by men.

1.8. The basal subjects

The *basal* category comprises subjects dying of violence, or natural causes having no known correlation with hypertension or atherosclerosis. The *basal* group offers an approximation to a representative sample of the population [33,34].

1.9. Subjects and methods

Methods summarized here are described more fully elsewhere [54]. Formalin fixed specimens of coronary artery and renal cortex were prepared at autopsy from men and women of black and white ethnic groups aged 15-79 years in the Orleans Parish Coroner's Office from 1992-98. Subjects of all cause of death categories were retained at ages 15-54 years, while only *basal* subjects were preferentially retained at ages 55-79 years. The proximal right coronary artery was opened longitudinally, cut into three segments of three cm length each, embedded in paraffin on edge to allow sectioning perpendicular to the luminal surface, and stained with hematoxylin and eosin (H&E). Renal cortex was sectioned in paraffin and stained with PAS-alcian blue. Atheromas such as in Figure 1-1E were marked in black ink on the cover slip of the coronary artery sections, and "fatty streak elements" such as in Figure 1-1C&D were marked in green ink. The percentage of specimen having atheroma (marked black) is called P_A and the percentage of nonatheromatous specimen with fatty streak elements (marked green) is called P_F. At nine equally spaced positions along the coronary sample, excluding black and green marked locations, measurements are taken of intimal thickness, counts are made of SMC's in a 100 μm wide band through the full intimal thickness, and the ratio of thickness to SMC's per 100 μm is calculated. Means of these three variables are called F, C_T, and F_C respectively. Intimal thickness of renal interlobular arteries is determined as percentage of outer diameter, the averages of these percentages in the size ranges of 80-149 μm and 150-300 μm are called R_r and R_c respectively, standing for renal resistance level and renal conduit level vessels. R_a is the mean of R_r and R_c. For later graphical purposes, the scale of measure of R_c is changed into 1.5xR_c so that the units of measure are μm of intimal thickness normalized to an artery of 150 μm diameter, and for R_r to1.0xR_r normalizing to a diameter of 100 μm.

Table 1-1 offers a description of the variables in two broad age ranges, and provides information for multivariate analysis such as correlation and regression.

1.10. A note of caution

The average number of intimal SMC's observed under a unit area of endothelium in a coronary artery increases slightly after intrusion of atheromas, as seen in the preserved regions of artery not censored by atheroma ($r = 0.20$ between YesA and C_T in Table 1-1). This effect is likely due to atheromas selectively occupying and destroying the arterial locations with the fewest intimal SMC's, a phenomenon that receives much attention in later chapters. Thus a protective action of SMC numbers serving to prevent atheroma can spuriously generate a paradoxical positive coefficient for C_T in Equation 1-3. The entry of C_T into Equations 1-1 and 1-2 could also be spurious for the same reason. This topic is elaborated especially in Chapters Four and Fifteen.

Table 1-1a. Description of variables including product moment correlation coefficients; all 272 cases aged 35-79 years.

Variable name	Symbol	Units	Mean	SD	Correlation coefficients †								
					F	F_C	R_a	Hy	C_T	P_F	S	B	YesA
Age	A	Years	49.3	11.0	.23**	.34**	.52**	.02	.05	.04	-.08	.04	.32
Fibroplasia													
Coronary total	F	μm	248	153	.---	.68**	.26**	.23**	.66**	.43**	.30**	.10	.57**
per SMC	F_C	μm	9.6	4.4	.66**	.---	.23**	.22**	.00	.39**	.20**	.00	.60**
Renal average	R_a	%od	9.2	5.1	.17**	.07	.---	.30**	.14*	.07	.04	.-.12	.28**
Hyaline													
Arterioles	Hy	$\sqrt{(N/cm^2)}$	2.8	2.4	.23**	.22**	.34**	.---	.11	.16**	.19**	.07	.28**
SMC numbers	C_T	SMC's	26.3	10.7	.67**	-.02	.14*	.11	.---	.21**	.26**	.22**	.20**
Foam cells	P_F	%	15.5	16.3	.43**	.40**	.06	.16**	.21**	.---	.22**	.00	.44**
Sex	S	M=2,F=1	1.30	0.46	.33**	.24**	.09	.19**	.26**	.23**	.---	-.03	.31**
Race	B	B=2,W=1	1.50	0.50	.09	-.01	-.16**	.07	.22**	.00	-.03	.---	.17**
Atheroma	YesA	N=0,Y=1	0.57	0.50	.54**	.55**	.15*	.29**	.20**	.46**	.35**	.16**	.---

† Raw correlations in upper right, age adjusted partial correlations in lower left.
* Significantly different from zero, $p < .05$; ** = $p < 0.01$.

Table 1-1b. Description of variables including product moment correlation coefficients; all 98 cases aged15-29 years.

Variable name	Symbol	Units	Mean	SD	Correlation coefficients†					
					F	F_C	R_a	Hy	C_T	P_F
Age	A	Years	22.7	4.1	.33**	.19	.33**	.48**	.24*	.00
Fibroplasia										
Coronary total	F	µm	94	47	.---	.36**	.34**	.06	.89**	.26
per SMC	F_C	µm	5.3	1.2	.36**	.---	.19	.14	-.08	.15
Renal average	R_a	%od	2.7	1.9	.26**	.14	.---	.37**	.26**	.23*
Hyaline										
Arterioles	Hy	$\sqrt{(N/cm^2)}$	0.7	0.9	-.12	.06	.25**	.---	.00	-.08
SMC numbers	C_T	SMC's	17.9	8.6	.89**	-.13	.20*	-.13	.---	.22*
Foam cells	P_F	%	8.8	13.6	.28**	.16	.24*	-.09	.23*	.---

† Raw correlations in upper right, age adjusted partial correlations in lower left.
* Significantly different from zero, $p < .05$; ** = $p < 0.01$.

CHAPTER TWO

THE SAMPLING THEORY OF FIBROPLASTIC ARTERIOSCLEROSIS

Abstract. The microscopic study of tissue sections requires sampling of whole arteries. The lateral walls of human thoracic aorta provide the necessary random replicate samples to support the appropriate use of sampling theory.

The histological study of human arteries requires sampling, and this in turn calls for the use of sampling theory to achieve meaningful inferences from the study of samples. Yet the arterial tree does not readily lend itself to the requirements of sampling theory because replicate random samples cannot be drawn from locations in the arterial tree. This is because the tree is not a level playing field; each location is unique and can be sampled only once. Fortunately for the purposes of this chapter, the locations along the lateral walls of the thoracic aorta are atypical and have the unusual

Table 2-1. Ratios* of fibroplastic intimal thickness at each locus to the average thickness of all loci, averaged at each locus over 121 NoA cases.†

	Right		Left			Right		Left			Right		Left	
Intercostal	Locus	Ratio	Locus	Ratio	Intercostal	Locus	Ratio	Locus	Ratio	Intercostal	Locus	Ratio	Locus	Ratio
4	1	1.11	51	0.93	.	18	1.02	68	1.04	.	35	0.92	85	1.00
.	2	1.10	52	0.97	.	19	1.00	69	0.99	.	36	0.92	86	0.96
5	3	1.10	53	0.99	8	20	1.01	70	1.05	.	37	0.88	87	0.95
.	4	1.07	54	1.03	.	21	0.98	71	1.04	.	38	0.88	88	0.97
.	5	1.06	55	1.09	.	22	0.97	72	1.08	.	39	0.88	89	1.00
.	6	1.03	56	1.12	.	23	1.00	73	1.06	.	40	0.87	90	1.00
.	7	1.04	57	1.13	.	24	0.99	74	1.05	.	41	0.87	91	1.04
6	8	1.03	58	1.14	.	25	0.94	75	1.06	11	42	0.90	92	0.99
.	9	1.02	59	1.19	9	26	0.91	76	1.05	.	43	0.90	93	0.96
.	10	1.08	60	1.13	.	27	0.89	77	1.01	.	44	0.90	94	0.98
.	11	1.06	61	1.13	.	28	0.90	78	1.02	.	45	0.83	95	0.94
.	12	1.07	62	1.10	.	29	0.92	79	0.98	.	46	0.78	96	0.96
7	13	1.06	63	1.11	.	30	0.94	80	0.99	.	47	0.78	97	0.98
.	14	1.03	64	1.09	.	31	0.94	81	1.02	.	48	0.78	98	0.99
.	15	1.04	65	1.12	.	32	0.91	82	0.99	.	49	0.78	99	0.96
.	16	1.05	66	1.12	.	33	0.95	83	1.01	12	50	0.79	100	0.96
.	17	1.01	67	1.07	10	34	0.95	84	1.02					

*By ANOVA these ratios do not differ significantly between age groups.

† Reproduced from Reference [73] with permission from Lippincott Williams & Wilkins

property of forming a nearly homogeneous population, insofar as they manifest arteriosclerosis. Samples of lateral thoracic aorta from the 4th to 12th intercostal levels, and especially from the 6th to the 10th levels, demonstrate this property as follows:

Excise the lateral walls, cut these into 2 cm segments, embed these in paraffin on edge to permit sectioning perpendicular to the luminal surface, stain with a connective tissue stain (for later purposes H&E is preferred), mark the tissue sections at 50 equally spaced locations on the left and 50 on the right. Discard all cases that reveal an instance of atheroma (Figure 1-1E) retaining only NoA cases. At the marked locations measure the intimal thickness, F, which in NoA specimens is by definition *fibroplastic* intimal thickness.

2.1. Locus bias

For each aorta, find the average of F over the 100 observed locations, and divide this into F_i at each location, i = 1 to 100, to express F as a ratio to the mean of F. For each position, i, average the ratios over all cases. Table 2-1 reports the results of this procedure in 121 aortas representing men and women of ages 15-69 years. The ratios detailed in Table 2-1 range closely around 1.0 with a standard deviation of 0.093. With no serious exceptions, the high ratios fell near the aortic arch, with ratios gradually decreasing toward a minimum at the 12th (subcostal) ostia; and the ratios were generally higher on the left than on the right. For very precise work, the tabulated ratios can be used to adjust the data in each case to compensate for this documented locus bias. In practice, however, the "noise" implied by a coefficient of variation of 9% seldom calls for this extra step. Rather, this source of noise can be judged as negligible, and samples taken along the lateral thoracic aorta can be treated as random replicates representing the whole population of loci. This seems especially reasonable if the sample is drawn between the 6th and the 10th intercostals.

2.2. Plaque-like clustering

Lesions of atherosclerosis are known to be plaque-like so that thickness of the intima at a locus is apt to resemble the adjacent and nearby loci. This matter was examined as follows: For each case, the 50 measurements of intimal fibroplasia along the right lateral wall were divided by the ratios in Table 2-1 to remove locus bias. Call the adjusted thicknesses X_i, where i is locus ranging from 1 to 50. With $x_i = X_i - M$, where M is the mean of the 50 values of X_i, compute $r = \Sigma x_i x_{i+1} \div \sqrt{(\Sigma x^2_i \, \Sigma x^2_{i+1})}$ where i ranges from 1 to 49. The 121 values of variances and covariance pooled over all cases were used to calculate the correlation shown in Table 2-2, r = 0.654. With $r = \Sigma x_i x_{i+2} \div \sqrt{(\Sigma x^2_i \, \Sigma x^2_{i+2})}$, i = 1 to 48, r = 0.480 was determined, and so on as tabulated. When loci exceeded 4 units of separation (approximately 1.3 cm in most aortas) their R^2 was less than 5 % ($r \leq 0.212 = \sqrt{0.045}$), indicating that plaque like clustering beyond 1.3 cm is generally insignificant. Hence, samples of loci taken at greater separations than 1.3

Table 2-2. Plaque like clustering of fibroplastic thickness revealed by correlation of points separated by 1, 2 . . . 20 units in 121 NoA cases.†

Units of separation	Correlation	Units of separation	Correlation
1	0.654	11	-0.015
2	0.480	12	-0.044
3	0.364	13	-0.062
4	0.283	14	-0.051
5	0.212	15	-0.053
6	0.150	16	-0.057
7	0.114	17	-0.043
8	0.070	18	-0.040
9	0.046	19	-0.031
10	0.016	20	-0.026

† Reproduced from Reference [73] with permission from Lippincott Williams & Wilkins

cm can be treated as independent for many practical purposes.

2.3. Distribution of fibroplasia over the lateral thoracic aorta

Table 2-3 reports the frequency distributions of the variable F in four representative cases. These were chosen to exemplify high and low standard deviations in company with high and low means. In every case the distribution is unimodal with an upward skew, and this was true for all 121 NoA cases. This outcome has one consequence of immediate importance. The "fibrous plaques" recognized by gross inspection of these aortas are not qualitatively different from the so called "normal intima", but are simply the upper tail of a continuous distribution in these NoA specimens. This result is illustrated in Figure 2-1.

Aortas A, B, and C in Figure 2-1 illustrate the manner of sampling for histology. The thoracic aorta is opened longitudinally, flattened, and fixed in formalin. (These examples also were stained with Sudan IV, but this is inconsequential here.) The excised lateral walls are prepared for histology. The excised samples in these three instances all happen to include examples of "fibrous plaques", white arrows, to be contrasted with "normal intima", black arrows. The histologic appearances of these six sites are also shown.

The "normal intima" of specimen C has greater fibroplastic intimal thickening

Table 2-3. Numbers of loci having fibroplastic thickenings of specified degree in four representative cases.

Thickness; μm	Case a	Case b	Thickness; μm	Case c	Case d
21-40		1	101-140		2
41-60		7	141-180		5
61-80	9	12	181-220	4	18
81-100	13	14	221-260	10	25
101-120	20	11	261-300	21	10
121-140	11	14	301-340	18	10
141-160	19	16	341-380	13	5
161-180	11	11	381-420	10	4
181-200	11	4	421-460	7	5
201-240	3	3	461-500	6	4
221-240	1	1	501-540	4	0
241-260	2	2	541-580	4	3
261-280		0	581-620	3	3
281-300		1	621-660		2
301-320		0	661-700		3
321-340		1	701-740		1
341-360		2			

† Reproduced from Reference [73]; with permission from Lippincott Williams & Wilkins

than the "fibrous plaque" of specimen B. The "normal intima" of specimen B has greater fibroplasia than the "fibrous plaque" of specimen A. These three aortas all obeyed the same skewed bell shape as in Table 2-3, and the photographs depicting "fibrous plaque" serve to illustrate the upper tails of those distribution curves. (Note: In another widely employed usage, the term "fibrous plaque" refers to atheromas such as in Figure 1-1E. The word "fibrous plaque" has multiple disparate meanings, a commonplace quality of the English language. This complexity should not distract us here. When used in the sense employed here "fibrous plaque" is defined by the appearance viewed in the gross by the unaided eye, irrespective of what the later histology might show. More often than not, the grossly defined fibrous plaque does not reveal atheroma in later histology.)

The mathematical form of the distributions in Table 2-3 is the negative binomial [66,74], a form that is closely related to the Poisson. It is the form that describes many kinds of scattering, such as the tossing of pennies onto a checker board. To follow this reasoning, suppose that fibroplastic arteriosclerosis is assembled from units that are about the size of a smooth muscle cell (SMC) and its associated

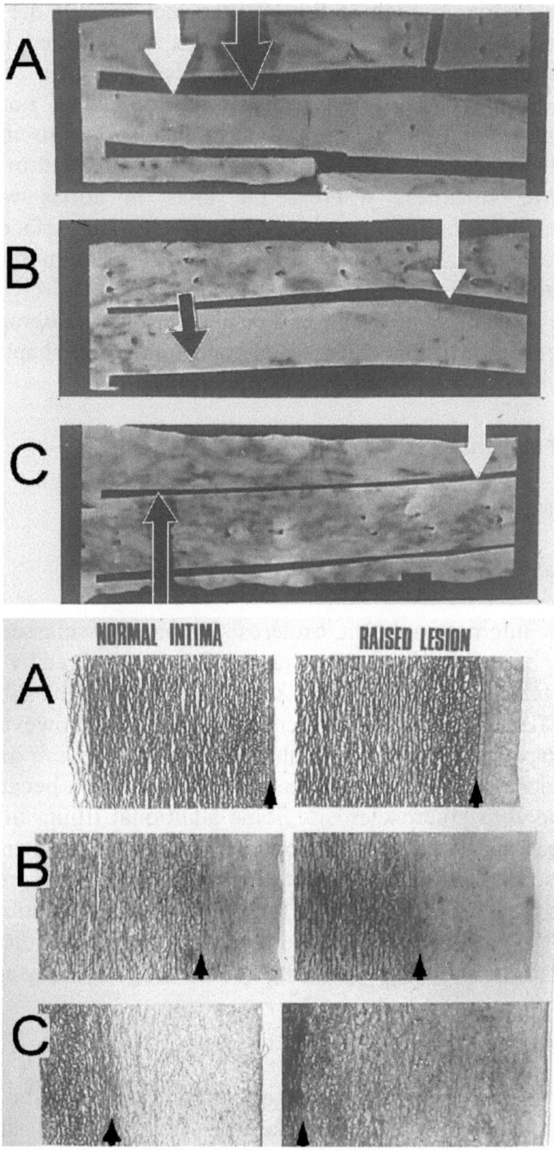

Figure 2-1. Three males aged 45-54 years are presented. The fixed and Sudan stained gross specimens (above) are sampled by excising the lateral walls of the thoracic aortas. Within this sample can often be found instances of "fibrous plaques" (white arrows) as well as "normal intimal" (black arrows), grossly defined, and their histology is shown in the photomicrographs (arrow heads mark intima-media boundary; trichrome, 70X). Reproduced from Reference [73] with permission from Lippincott Williams & Wilkins.

collagenous matrix. When these units are scattered over the sample space, the total bulk of the collagenous matrix in each of the affected loci follows the distributions as exemplified in Table 2-3. When the bulk is large, the affected location bears a fibroplastic thickening that is visible to the unaided eye as a "fibrous plaque". Otherwise the mildly undulating arterial surface will be called "normal intima", or sometimes "diffuse intimal thickening", a term that serves to introduce further confusion. The unit composing the normal intima was estimated in Chapter One at about 5.0 μm/SMC thickness whereas the units in aortas with fibroplastic arteriosclerosis were described as exceeding 9.3 μm/SMC. The surface undulations are exaggerated in the latter kind of aorta when viewing the intimal surface with the unaided eye. Visualizing this unit in histologic sections is not entirely straightforward, and calls for some conceptual preparation. For this reason, photographs showing the morphological appearance of this unit are postponed until later chapters.

2.4. Materials and methods

These are summarized from a previous report [74].

2.4.1 Case selection

Male subjects in the International Atherosclerosis Project were classed by five 10-year age groups and by six grades of severity of raised lesions as judged by gross inspection of the thoracic aorta. The intention was to select 10 cases in each subgroup, yielding 300 cases in all. Many of the subclasses could not be filled, however, because such cases were rare. Especially rare were young subjects with severe disease and older subjects with few lesions. Where possible, cases were chosen because the cause of death was not related to atherosclerosis. Some additional filling of subgroups was possible by using cases in which the cause of death was related to atherosclerosis (eight such cases were used). After choosing 230 cases in this way, 25 were later discarded because they were technically inadequate for evaluation or were subsequently found to have features of syphilitic aortitis, leaving 205 cases to be studied. Aortas in the archives had been formalin-fixed after being opened longitudinally and flattening.

2.4.2. Tissue processing

Samples of lateral wall were excised from the fourth intercostal to the 12th (subcostal) ostia, decalcified in formic acid, blocked in paraffin, and stained with an elastin-collagen trichrome procedure. Lines were drawn onto the cover slips so that the length of each lateral wall was divided into 49 equal segments to mark 50 points to be graded. Distances between points varied according aorta around an average of about 0.3 cm. The features measured in a projected image were (1) full intimal thickness in the absence of atheronecrosis, F_1, (2) thickness of the focus of atheronecrosis, A, (3) thickness of the fibroproliferative cap over the atheronecrosis, C, and (4) thickness of the fibroproliferative base underlying the atheronecrosis, B. These quantities are one-dimensional, representing distances perpendicular to the endothelial surface. Atheronecrosis was defined as a generally unstained (white) area containing clearly recognizable cholesterol clefts and black-stained debris. Measurements of thickness are subject to error from a number of identifiable sources. Artifactually there can be loss of superficial intima, compression of tissue during storage or handling, or fragmentation with falling apart of the intimal layers. Sometimes aberrantly situated intercostal ostia or extra ostia were included in the sample. The convention was adopted that, when any of these artifacts or aberrant ostia were identified at the point marked for evaluation, the nearest acceptable place on the specimen was substituted for the unsuitable one. It was not always possible to distinguish a clear boundary between media and intima. Other vagueness of boundary was also sometimes found. It is judged that these sources of error can be treated as unbiased insofar as the measurements are concerned. On the other hand, the locations affected by atheroma are not random with respect to the fibroplasia that they displace, and this source of sampling bias is of major concern throughout what follows. Of the 205 cases, 84 were classed as YesA, because atheroma was found in the sample; these are discussed in Chapter Three. The other 121 are subjects of Chapter Two.

CHAPTER THREE

INTRUSION OF ATHEROMA INTO THE MOST FIBROTICALLY THICKENED INTIMAL SITES.

Abstract. A large bulk of fibroplastic matrix around the intimal SMC's could be the usual rate limiting step in the evolution toward atheroma. Does that matrix act by way of trapping atherogenic lipids? Sampling theory can provide important evidence.

Atheromatous necrotic core can occur at any depth in the fibroplastic intima of the lateral thoracic aorta, with a weak tendency to favor a deep location. These conclusions were obtained as follows: Of the 205 subjects in this study, 84 cases revealed atheroma in some part of the histologic samples of aorta (cf. Article 2.4.1 in Chapter Two). A total of 1,135 points with atheroma were encountered in the 84 cases combined. The most frequently encountered thickness of the necrotic core was about 350 μm. Each point with atheroma had a fibrous cap over the necrotic core and a fibrous base under it in most instances. The coexistence of thin cap with thick base is illustrated by Frame B in Figure 3-1, thick cap with thin base by Frame D, and equal cap and base by Frame C. Absence of the cap (ulcerated plaque) is illustrated by Frame A and absence of base by Frame E. The relative percentages of the patterns A, B, C, D, and E were respectively 6.7, 20.0, 22.5, 39.8 and 11.0 %. It seems likely that the rupture of the thin cap is apt to be from postmortem autolysis in the 6.7 % of atheromas with this feature, because premortem rupture is expected to precipitate thrombosis and this is rarely seen in these examples. Similarly, medial erosion after obliteration of the fibrous base was uncommon in these specimens.

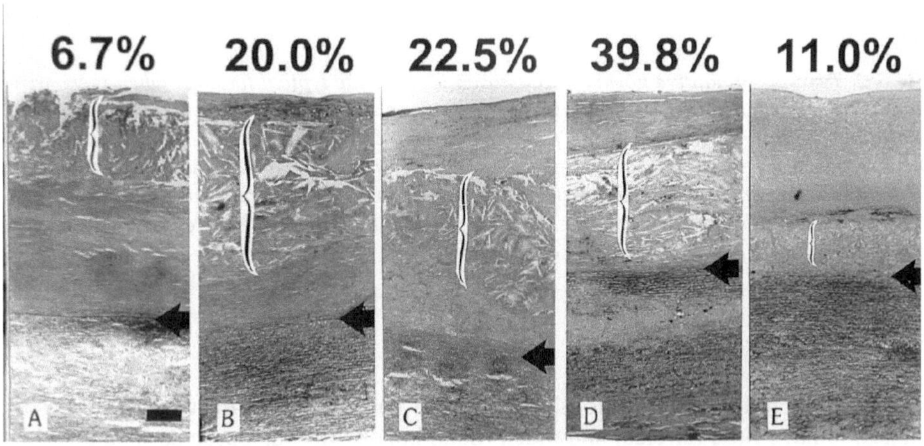

Fig 3-1. Examples of atheromas were chosen to illustrate necrotic core at various depths between luminal surface and intima-media boundary (arrows). Necrotic core (brackets) is readily recognized by finding crystals of cholesterol, easily seen in these views. Nigrosin-trichrome paraffin sections, bar = 200 μm.

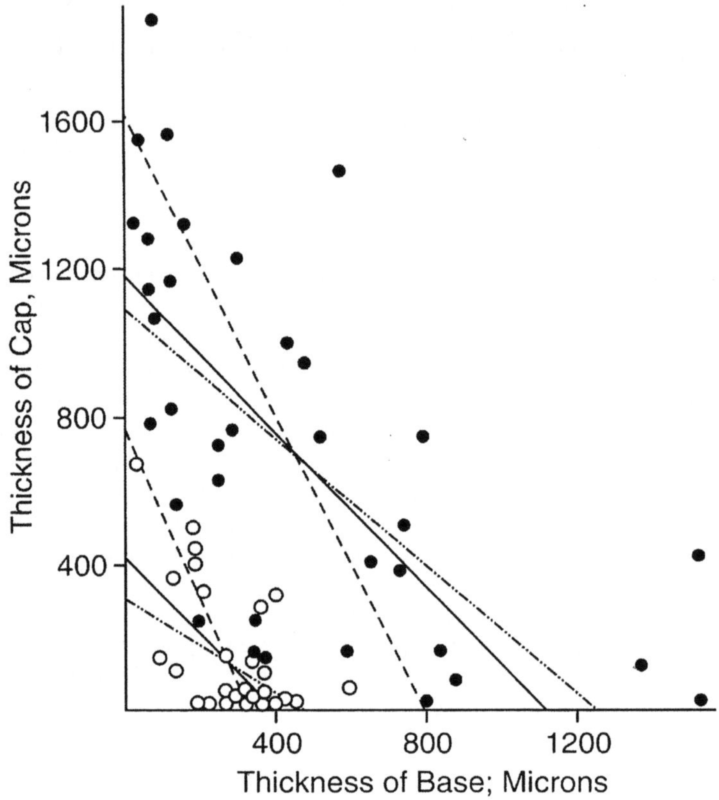

Fig 3-2. Two representative aortas were chosen to illustrate typical findings when there was little (circles) or much (dots) fibroplasia in the sum cap + base. Each symbol within an aorta represents a pair of measurements from a different and independent lesion. Sum of the measurements, F_2 = cap + base, averaged 425 and 1164 μm respectively, and the cases ranked 81th and 10th on this variable. Regression lines of cap on base and of base on cap are sketched (broken lines). Solid lines have slope -1 for comparison. Reproduced from reference [72] with permission from Lippincott, Williams & Wilkins.

3.1. Relation of cap to base

Figure 3-2 diagrams the relationship of cap to base observed in two typical cases chosen to have little (circles) and much (dots) bulk of fibrous tissue associated with atheroma. A negative correlation, as in these two examples, was found in 72 of the 84 cases, and was of statistical significance ($p < 0.05$) in 32 of them. In the pooled covariances derived by fitting parallel lines to the 84 cases, the overall correlation between cap and base was -0.383. A slope of -1 (solid lines in Fig. 3-2) fell within the

95 per cent confidence limits of the statistically computed slopes (linear regression) in all cases. The conceptually simple slope of -1 is therefore supported by the data. This finding implies that the depth of intrusion by the atheroma does not affect the sum of cap + base = F_2 in any systematic way. The sum tends to fluctuate around a constant value that defines a general property of each aorta. The quantity F_2 is the *fibroplastic thickness* as opposed to the *atheromatous thickness* at each observed site, and measures a constituent of intima comparable to F_1 at nonnecrotic sites, and to all sites in NoA arteries.

3.2. Relative depth of atheroma

The mean of F_2 fell in the range 600 to 1300 μm in 90 % of cases. The mean ± standard deviation of the 84 cases was 868±240. Expressing the cap as a percentage of F_2 yields P = 100cap/F_2. P ranged from 36 to 80 % in 90 % of cases. Mean±SD of P was 58.7±14.9 %. The tendency for the cap to exceed 50 % of the fibroplastic thickness was small. This finding reinforces the percentages in Figure 3-1, implying that the atheroma shows no great proclivity to intrude at any particular depth in the fibroplastically thickened intima. Knowing that oxygen tension decreases with distance from lumen surface, this finding argues that oxygen tension bears little if any influence on intrusion by atheroma.

3.3. Erosion of the fibrous cap by atheroma

The fibroplastic component of intimal thickness, F_2, tended to be large if the size of the necrotic core was small, and to decrease as the core grew large. This outcome was shown as follows: From the pool of 1,135 points among all 84 cases, points were grouped into intervals on atheroma size (1 to 100, 101 to 200. . . .). Means of cap, base, and F_2 for each of these groups of points are given in Figure 3-3. The average F_2 tended to decrease with increasing size of necrotic core in a proportion of 1:4 and tended toward 1000 μm at core = 0. Defining a new variable, $F_3 = F_2 + x*\text{atheroma}$, the coefficient x can be assigned the convenient value of 0.25, extracted graphically from Figure 3-3. The decrease in the sum, F_2 = cap + base, with enlarging core appears to be entirely in the cap, with the base remaining stable throughout all sizes of necrotic core. When F_2 and F_3 were averaged for each of the 84 cases, the correlation between the two averages was 0.981. Therefore, for characterizing each aorta, the quantity F_3 adds no important information to the more easily obtained quantity F_2. For the construction of a later table, however, F_3, seems more appropriate. The data offer little support for the proposal that the necrotic core may induce formation of a thick fibrous cap. Such formation should act to impart high values to P, an action that could account for mean P in excess of 50 %, but the observed value of 58.7 % offers only small support for such an action. Rather, an erosion of the cap by expanding core is suggested by the findings in Figure 3-3.

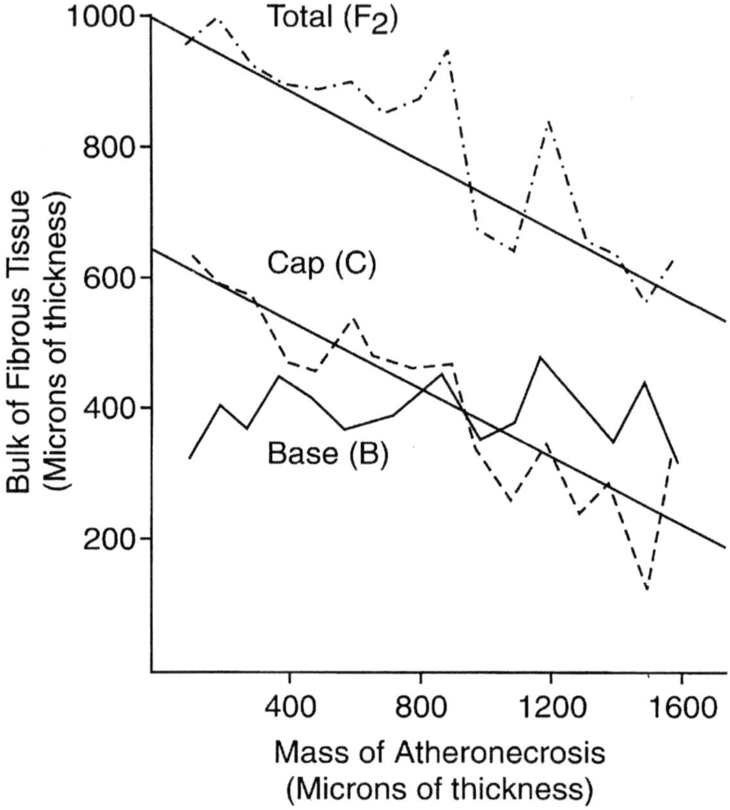

Figure 3-3. Thickness of atheroma was used to group the points into intervals (1 to 99, 100 to 199 . . .). Means of fibroplastic thickness associated with the necrotic core are plotted. Solid lines with slope -0.25 are sketched for comparison. Reproduced from reference [72] with permission from Lippincott, Williams & Wilkins.

3.4. Atheroma favors aortas with the greatest generalized fibroplastic intimal thickenings throughout the specimen

Figure 1-3 in Chapter One illustrates this matter for coronary arteries. Arteries with thickened intima throughout the specimen often show atheromas at some locations. When W exceeds zero in the following equation, derived from the total pool of 205 aortas, the probability of finding atheroma exceeds 50 %.

$$W = 0.84F_1A \quad (T^2 = 1.82 \quad \%=77 \quad c=-2.4) \qquad (3\text{-}1)$$

A is age in years, F_1 is mean fibroplastic intimal thickness in μm at sites without atheroma, and other symbols are explained in Article 1.7, Chapter One. (A similar replicate equation also derived for another set of aortas [60] was found to be W = $0.63F_1A$ (T^2 = 3.80 %=85 c=-7.7) in good agreement with Equation 3-1.) The coronary and aortic data sets, summarized in Equations 1-3 and 3-1, agree closely with each other on the important principle: Atheroma has a strong tendency to appear in arteries with widespread fibroplastic intimal thickenings as a general feature throughout the specimen. If atheroma is in some way causing the fibroplasia, then we must explain how this happens at sites many centimeters away from the atheroma itself. A much easier explanation of these findings is that fibroplasia attracts the atheroma by acting as a lipid trap.

The term for age in Equation 1-3 became unnecessary when SMC numbers enter into consideration. A similar analysis could not be pursued here with these aortas, because the data set under consideration here did not include SMC enumerations.

3.5. Atheroma favors the greatest fibroplastic thickenings

Points having and not having atheroma formed two non overlapping sets of data; each of the 100 points in a case was one or the other but not both. The sum of cap plus base held mean values greater than mean values for the total intimal thickness at sites without atheroma, i.e. $F_2 > F_1$, in 76 of the 84 cases. In 81 of the 84 cases the mean adjusted fibroplasia represented by F_3 exceeded the mean of F_1. The averages of F_2 and F_1 respectively for the 84 cases were 868 and 598 μm of thickness. Clearly, the atheroma tended to be found at the intimal points with the greatest bulk of fibroproliferative intimal thickness, and this tendency prevailed in nearly all cases whether mean F_1 was large or small. Atheroma, therefore, tended to reside in those points in a case that were the bulkiest in intimal fibroplasia.

Over the 84 cases, a correlation of 0.395 was found between F_2 and F_1. Between F_1 and number of points having atheroma (n), the correlation was 0.514. (Mean n was 13.5 and the correlation between n and F_2 was 0.093.) The two significant correlations, 0.514 and 0.395, indicate that atheroma were most numerous in cases with the general property of having the bulkiest intima even at sites remote from atheromas, thus reinforcing the conclusions from Equation 3-1. Hence atheroma tends to favor the greatest fibroplastic thickenings between cases as well as within cases.

3.6. Frequency of atheroma in relation to general intimal fibroplasia

Let us introduce the term general fibroplasia to subsume F_1 or F_3 indiscriminately. Alternatively it could apply to F_1 or F_2 as appropriate. The first choice is preferred here to avoid distortions imposed by a few outlier values with atheromas of extraordinary size. The 8400 data points from the 84 YesA cases were grouped into intervals on

Table 3-1. Percentage of atheromas with increasing fibroplastic intimal thickness in 8400 sites examined in 84 YesA aortas.†

F_1 or F_3	Atheroma %	No. of sites
0-99	0.0	24
100-199	0.0	371
200-299	0.0	1220
300-399	0.8	1228
400-499	2.8	922
500-599	9.8	798
600-699	14.2	726
700-799	20.6	612
800-899	25.1	541
900-999	30.4	405
1000-1099	34.5	377
1100-1199	34.0	318
1200-1299	31.7	243
1300+	37.4	615

† Reproduced from reference [72] with permission from Lippincott, Williams & Wilkins

general fibroplasia. The percentages of points having atheroma in each interval are given in Table 3-1. Atheroma was never observed with general fibroplasia less that 300 μm. The percentage of points with atheroma rose from 0.8 % to 34.0 %as fibroplasia increased from 300 to 1100 μm, and then remained stable with further increases in fibroplasia. These results further emphasize the conclusion that atheroma favors intimal sites with the greatest fibroplastic thicknesses.

3.7. Similar results in the coronary arteries

Findings in the coronary arteries of 173 cases [61] revealed patterns much like those for the aorta. Unlike the lateral thoracic aorta, of course, the coronaries do not offer a level playing field, and many assumptions must be made concerning the statistical validity of the tissue sampling methods. Those assumptions seem justified for some limited purposes after seeing the validation of them for the aorta in Chapter Two. The similarity of outcomes here for the aorta and coronary arteries indicates that the strength of this effect can emerge even when sampling biases are substantial.

Each of the 173 cases was evaluated at 27 sites separated by 1 cm along the

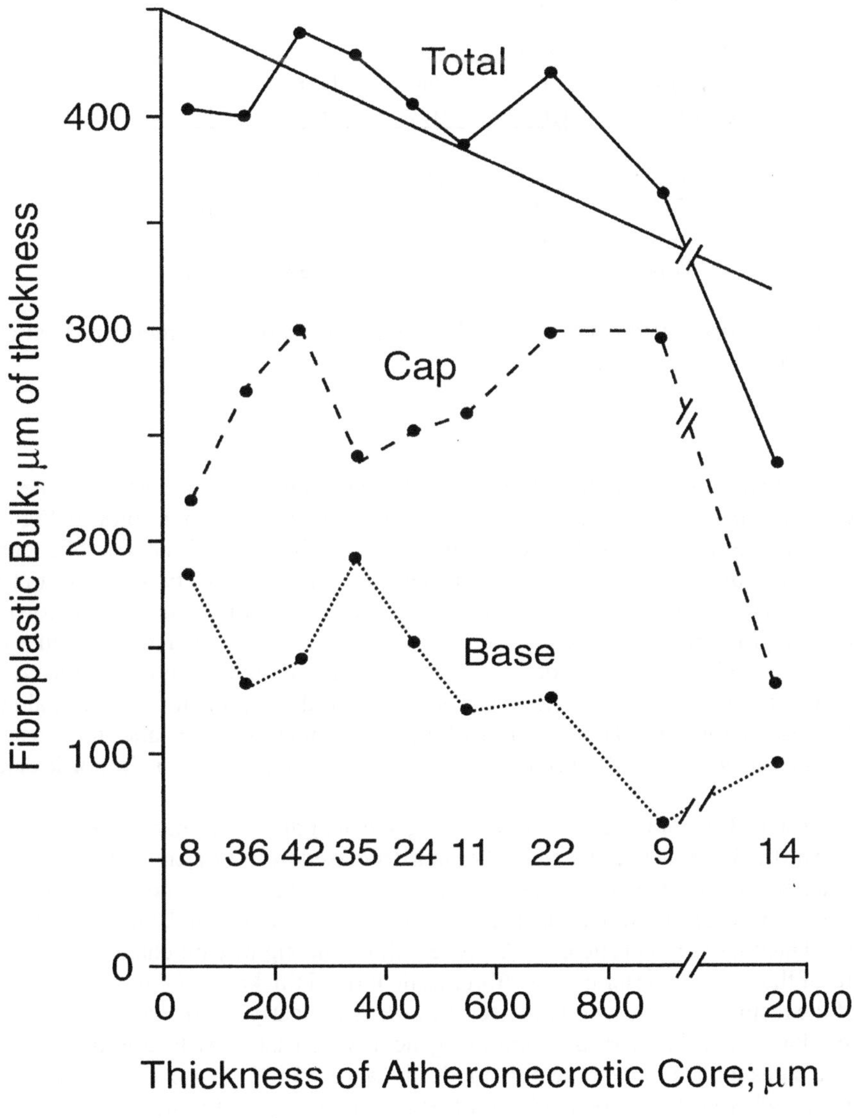

Figure 3-4. As in Figure 3-3; data for coronary arteries. Solid line with slope -0.125 is sketched for comparison. Reproduced from reference [61] with permission from the American Medical Association.

Table 3-2. Mean thickness of necrotic core, cap and base in atheromatous sites by distance along the coronary arteries.†

Distance from origin; cm	Necrotic core µm	Cap µm	Base µm	N
0-2	394	281	146	90
3-5	438	253	134	76
6-8	406	236	102	36
Total	413	262	134	202

† Reproduced from reference [61] with permission from the American Medical Association.

first 8 cm of anterior descending, left circumflex, and right coronary arteries. The 52 YesA cases revealed atheromas at a total of 202 sites, and these form the plots in Figure 3-4, which reiterates the pattern of Figure 3-3. As the size of the necrotic core increases, the coronary data show an average decline in the fibrous base but not in the fibrous cap. The quantity F_2 decreases at about one eighth the rate of the core's growth, extrapolating to a value of 454 µm at zero core size. This result is tentative because of small numbers of data points, especially in the critical region near zero necrotic core, represented by only eight observations. The failure of the cap to show erosion along with the base might suggest proliferation of the cap in some sites to balance the erosion at other sites, in contrast to the absence of evidence for such a process in the aortic data.

Table 3-2 compares locations along the length of coronary arteries. The fibroplastic thickness at sites lacking atheroma decreased from 170 µm in the first third of the length to 102 µm in the last third, a 40 % decline (data not shown). The fibroplastic base of atheromatous sites shows a 30 % decline, and the cap a 16 % decline. Thickness of necrotic core, however, shows no significant change along the lengths of the coronary arteries. The fibroplastic intimal thickness seems scaled to the size of the resident artery, while the necrotic core is of nearly fixed size in all sizes of arteries. Published data for the aorta are generally consistent with this trend [60]. Necrotic cores tended to average only modestly larger (56 %) in the aorta than in the coronaries, whereas the associated solid tissue is markedly more bulky in the aortas (165%).

As in the aorta, the probability of finding atheroma increases with expansion of fibroplastic intima in the coronary arteries (Table 3-3). This confirms for the coronaries the important feature of aortas: atheroma tends to favor the greatest fibroplastic thickenings, and this occurs within cases as well as between cases.

Table 3-3. Percent of sites with atheroma by thickness of generalized fibroplasia according to distance along the coronary arteries in 52 YesA arteries.†

Generalized Fibroplasia F_1 or F_3 μm	Distance along Coronary; cm				Number of Sites
	0-2	3-5	6-8	Total	
0-99	0	0	0	0	352
100-199	10	7	4	6.7	313
200-299	25	13	10	15.7	204
300-399	33	29	31	31.2	138
400-499	36	28	29	31.2	109
500-599	38	43	20	37.5	64
600-699	62	40	17	46.8	47
700+	58	47	29	47.3	55

† Reproduced from reference [61] with permission from the American Medical Association.

CHAPTER FOUR

CONDITIONS FOR INTRUSION OF ATHEROMA IN CORONARY ARTERY

Abstract. Data on aorta in Chapter Three indicate that intimal fibroplasia may precede and dispose to intrusion by atheroma. Evidence in this chapter points to excessive matrix per SMC as the main impulse toward fatty degenerations, both fatty streak elements and atheroma. Furthermore, the evidence argues that the augmenting of fibroplasia is how aging acts to prepare for atheroma.

The observable sites within YesA arteries, after exclusion of sites censored by atheromas, are usually found to have generalized fibroplastic intimal thickenings throughout the specimen as a whole when compared with NoA arteries. The comparison is done in this chapter by discriminant function analysis, restricting the ages to 35-79 years, because the YesA condition is rare before age 35 years (cf. Chapter One, Article 1.9 for a description of the cases). The objective is to determine an equation in the form of Equation 4-1 which, when set equal to zero, maximally distinguishes the YesA from the NoA specimens, where each of the terms in X refers to a selected variable of interest. Descriptions of these variables can be found in Chapter One, Tables 1-1a&b.

$$Y = b_1X_1 + b_2X_2 + b_3X_3 + \ldots + a \tag{4-1}$$

An early stage of analysis discards age from the optimal form of Equation 4-1. The YesA cases are older than the NoA cases on average, but the equations that emerge from this analysis exclude age as not of statistical significance, because the retained X variables stand in its place. This finding raises the possibility that age acts by way of the retained variables, which implies that we know what age does to the tissues in preparation for the intrusion of atheroma.

4.1. Let $X_1 = F$, the fibroplastic intimal thickness pervading the coronary artery

Equation 4-2 is the discriminant function describing the pool of all 285 subjects aged 35-79 years.

$$Y = 1.71F \quad (T^2 = 1.96 \quad \%=79 \quad c=-2.65) \tag{4-2}$$

Symbols are explained in Chapter One, Article 1.7. The standardized coefficients are used in this chapter. They can be restored to raw form using standard deviations in Table 1-1a.

4.2. Let $X_2 = C_T$, total number of SMC's averaged over nine sites

Figure 4-1offers a graphic guide to the following development. If we know both

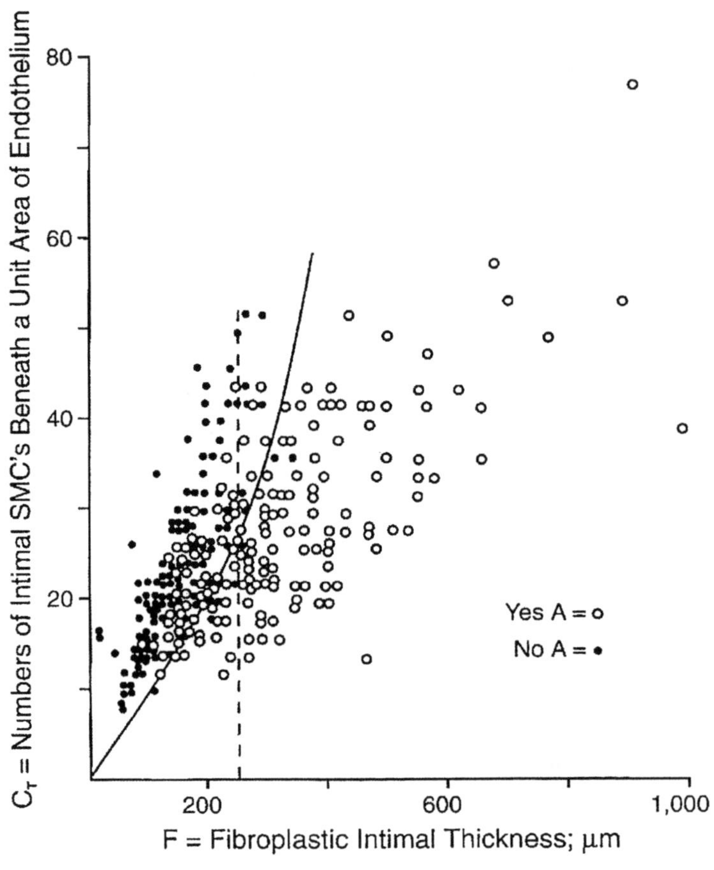

Figure 4-1. Men and women aged 35-79 years with all causes of death are represented. Vertical dashed line depicts Equation 4-2 with Y = 0. Curve depicts Equation 4-5 with Y = 0, which optimally distinguishes circles from dots.

quantities, intimal thickness, F, and numbers of SMC's associated with that thickness, C_T, then we can inquire about the relative contributions of each variable to predicting the YesA condition of each specimen. The vertical dashed line in the graph represents Equation 4-2 with Y = 0. Clearly the dashed line does not maximally separate circles from dots, showing that C_T adds useful information beyond that contained in F alone.

Let $X_3 = F_C$, fibroplasia per SMC

Equation 4-3 captures some of the information added by C_T, where F_C, the average of F/C_T in each specimen, represents the magnitude of fibroplasia per SMC.

$$Y = 1.89F_C \quad (T^2 = 2.30 \quad \% = 78 \quad c = -3.94) \tag{4-3}$$

Equation 4-3, if plotted in Figure 4-1, would describe a diagonal extending from the origin of the graph (and passing through the point 25.3,234 to be discussed later). Such a diagonal still does not maximally separate circles from dots, because further improvement is possible.

4.3. Combining X_2 and X_3

This relationship was summarized in Chapter One (Figure 1-3).

$$Y = 2.01F_C + 0.67C_T \quad (T^2 = 2.69 \quad \% = 82 \quad c = -5.81) \tag{4-4}$$

Equation 4-4 reproduces Equation 1-3, which was plotted in Chapter One separately for men and women as two nearly identical downward sloping lines in Figure 1-3 (the sex difference is not significant statistically). Symbols falling upward and rightward from the sloping lines are predicted to be YesA. Of the 127 YesA cases in the scatter plot for men, 94 have $Y > 0$ (74 %) and are correctly predicted this way. Symbols falling downward and leftward from the solid line are predicted to be NoA, and the 62 men with $Y < 0$ among the 64 NoA men (97 %) are correctly predicted in this way. Among women, 63% of YesA and 93 % of NoA subjects are correctly predicted by Equation 4-4.

Equation 4-4 identifies intimal fibroplasia as a likely precursor of atheroma, because the progression from youth to middle age begins near the left side of the graph in Figure 1-3 and spreads across the diagonal line into the acquisition of atheroma later in life. F_C dominates the relationship, with C_T offering a minor contribution, by 3 to 1 (2.01:0.67).

4.4. A note of caution

The average number of intimal SMC's observed under a unit area of endothelium in a coronary artery, C_T, is greater in YesA than in NoA specimens ($r = 0.20$ in Table 1-1). The small increase of C_T in atheromatous arteries is likely due to atheromas selectively occupying and destroying the locations with the fewest SMC's, as discussed in Chapters Five, Eight and Nine. Thus a protective action of SMC numbers serving to prevent atheroma can spuriously generate a paradoxical positive coefficient for C_T in Equation 4-4. The entry of C_T into other related equations could also be spurious for the same reason.

4.5. Combining X_1 and X_3 excludes X_2

Inclusion of both F and F_C jointly further improves results, while C_T is then rejected:

$$Y = 1.01F + 1.35F_C \quad (T^2 = 2.80 \quad \% = 81 \quad c=-\ 4.38) \qquad (4\text{-}5)$$

Equation 4-5 with Y = 0 is plotted as a curve in Figure 4-1. The curve crosses the vertical line near the point (25.3,235), which defines F_C = 9.3 µm/SMC, a quantity of much interest in Chapter One and in later chapters.

C_T, the average number of intimal SMC's beneath a unit area of endothelium, ranges among coronary arteries from 7 to 60 (with one outlier of 78) and averages 26.2 SMC's per sample in Figure 4-1. Some of this variation is due to the unmeasured factor of native arterial size, because the right coronary artery can be tiny or enormous from one heart to another. This and other sources of variation are nullified in the ratio $F_C = F/C_T$, leaving variations in C_T due to arterial size and other sources with no contribution to make in Equation 4-5.

The total intimal thickness, F, adds somewhat to the utility of F_C in predicting atheroma, and displaces C_T. These secondary effects are small, and could be, at least in part, spurious consequences of multicolinearities in these highly intercorrelated variables. Note that T^2 of 2.30 in Equation 4-3 rises only to 2.80 in Equation 4-5, a small change that could be considered negligible for some purposes.

4.6. Atheroma in 46 subjects without excessive F_C, the false negatives

In Figure 4-1, 46 open symbols fall leftward from the plotted curve representing Equation 4-5. These "false negative" YesA arteries are falsely classified as NoA by the discriminant function; they are of special interest. Among the aging coronary arteries, some of them failed to develop fibroplasia per SMC in excess of the threshold for diagnosing fibroplastic arteriosclerosis (those leftward of the curve). Nevertheless, atheroma did emerge in many of these arteries. On further inquiry these 46 false negative cases were generally found to contain very extensive involvement with foam cell infiltration of the kind illustrated in Figures 1-1C&D. This is shown by measuring P_F, the percentage of specimen infiltrated by foam cells (omitting sites with atheroma in YesA specimens as usual).

Means of P_F in the "false negative" YesA specimens are compared with the true negative NoA specimens in Figure 4-2. A trend to high P_F is seen in YesA specimens of all age groups, and is large and strongly significant at ages 35-54 years. Fatty streak elements can settle into these arteries even without the usual lipid trapping preconditions of high F_C seen in the ordinary atheromatous human subjects who are represented by the circles rightward of the curve. This is what happens in experimental animals such as cholesterol fed rabbits; its occurrence in human arteries is heavily

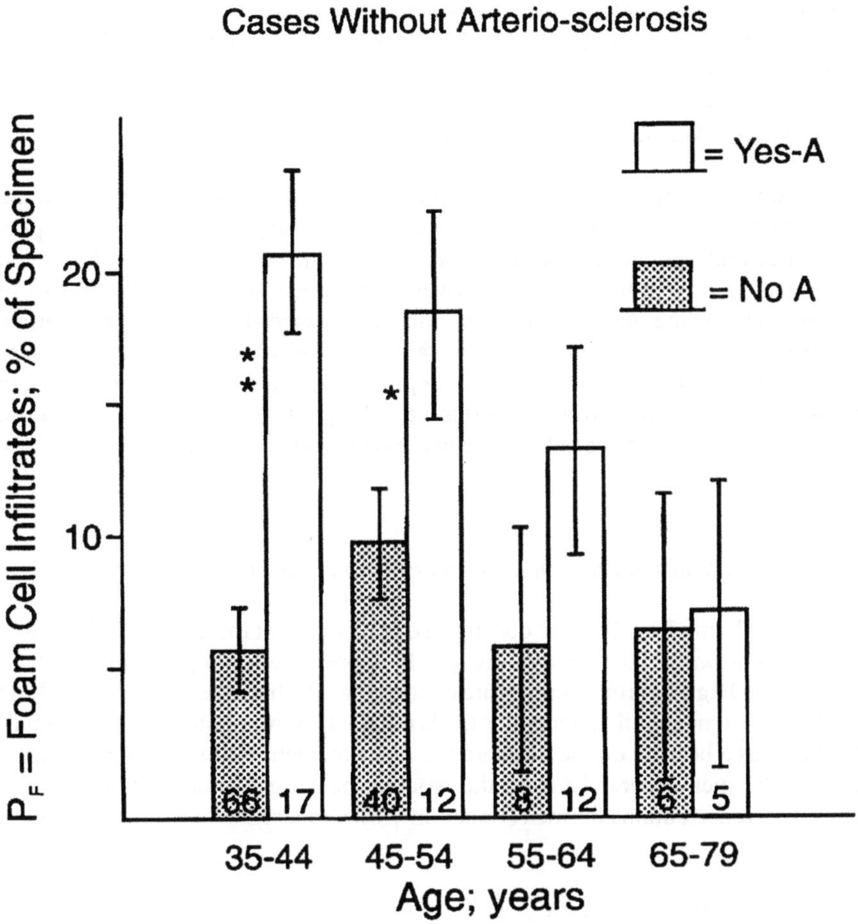

*Figure 4-2. YesA subjects plotted as circles leftward of the curve in Figure 4-1 (false negatives) are compared here with all NoA subjects. Line segments atop the bars represent standard errors of the means; ** and * represent p<0.01 and p<0.05 respectively in comparing adjacent columns. Numbers of cases at bases of columns.*

emphasized by the AHA Lesions Committee [10,11]. It is easy to suggest that these might be the subjects with severe dyslipidemias and other unfavorable coronary risk factors.

4.7. Lack of atheroma in 13 subjects with excessive F_C, the false positives

In Figure 4-1, 119 open symbols fall rightward of the curve. These are the "true positive" YesA arteries. The mean extent of foam cells as a percentage of the nonnecrotic portion of the specimen, P_F, was 23.8 % in these 119 cases, the highest level obtained in any assembly of cases. This is the outcome to be expected if arteriosclerosis acts as a lipid trap to promote formation of foam cell infiltrates [45,78,81], again in keeping with the AHA Lesions Committee position [10,11].

By way of contrast, 13 of the symbols falling rightward from the curve are dots; these are the "false positive" NoA arteries. These "false positive" arteries display arteriosclerosis that has not yet produced atheroma. The mean extent of foam cell infiltration was 12.8 % in these 13 cases, over double the values in the total pool of NoA arteries. This is the outcome, as before, to be expected if arteriosclerotic intima favors formation of foam cell infiltrates. The failure to find atheroma in these 13 cases could, in theory, be a result of inadequate sampling; the atheroma might be found, at least in some of them, if the coronary arteries were sampled more fully. The other false negatives might occur by chance alone due to method error, since their numbers are so few.

4.8. Foam cell infiltrates in relation to coronary intimal fibroplasia

Groups of High P_F and Low P_F were constructed by division at the median of P_F within each of the usual *basal* 10 year age groups. It is presumed here, as again in later chapters, that the High P_F groups constitute cohorts that all share the property of falling above average in a manner that generally tends to "track" throughout life. The High P_F classification has a bearing on the measure F_C. The statement of this relationship calls for handling the men separately from the women, because a small but statistically significant difference obtains for men (Equation 4-6) compared with women (Equation 4-6a).

$$Y_P = 0.71F_C + 0.40C_T \quad (T^2 = 0.60 \quad \% = 67 \quad c=-2.59) \qquad (4\text{-}6)$$

$$Y_P = 0.77F_C + 0.64C_T \quad (T^2 = 0.56 \quad \% = 65 \quad c=-2.83) \qquad (4\text{-}6a)$$

For brevity, only the data on men are plotted in Chapter One, Figure 1-2; the plot for women is nearly identical to this one for men except that the units on the two axes are on a reduced scale for women, especially for the C_T axis. The pattern has a resemblance to the comparable pattern for atheroma in Figure 1-3. This resemblance supports the view that fibroplasia might promote both atheroma and fatty streak elements, in keeping with Allbutt's proposal that arteriosclerosis is prone to fatty degeneration.

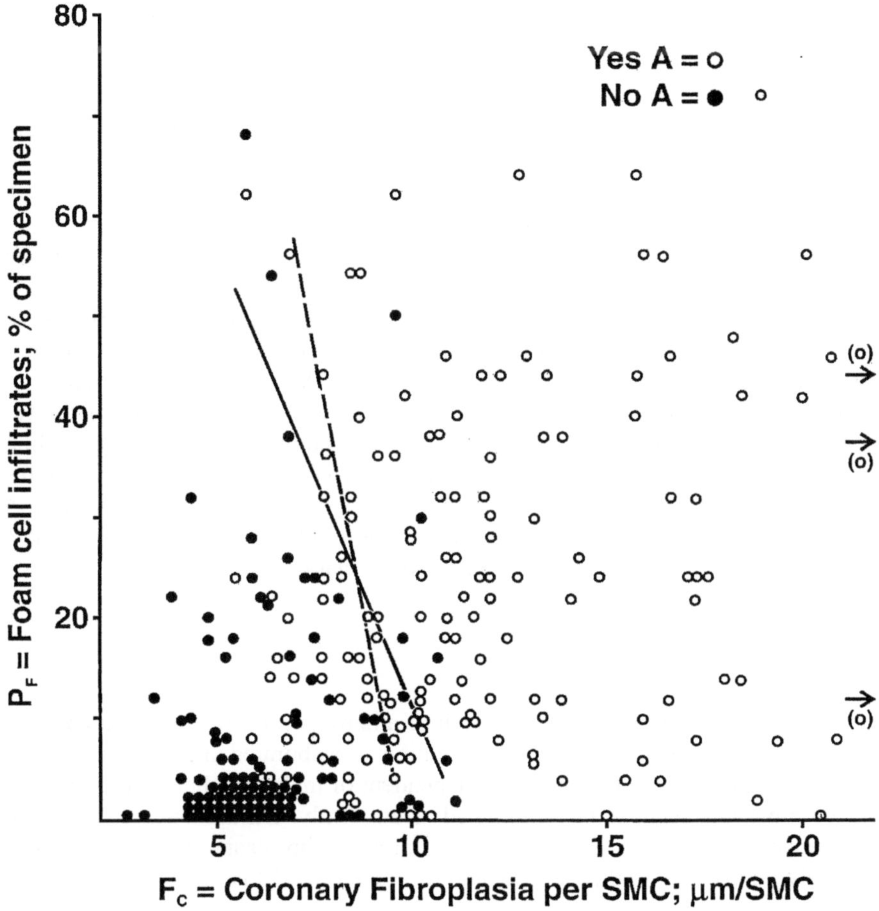

Figure 4-3. Symbols represent men and women indiscriminately. Nearly vertical lines represent optimum discrimination of circles from dots for men (solid) and women (dashed); these are Equations 4-7&7a with Y = 0. Off scale outlier values are noted by right pointing arrows.

4.9. Foam cell infiltration and the intrusion of atheroma

The measure of foam cell infiltration, P_F, is correlated with the measure of fibroplasia per SMC, F_C (age adjusted r = 0.40 in Table 1-1a). This correlation is visible in the scatter plot of Figure 4-3 as a tendency for symbols to move upward and rightward across the graph. The distance of each symbol away from the origin of the graph has an influence on the likelihood of finding atheroma in the specimen. This propensity of a coronary artery to become YesA can be predicted by expanding the variable list in Equation 4-3. These later equations all reject C_T as not of added significance.

$$Y = 1.39F_C + 0.72P_F \quad (T^2 = 2.24 \quad \% = 78 \quad c = -3.61) \tag{4-7}$$

$$Y = 3.29F_C + 0.96P_F \quad (T^2 = 4.86 \quad \% = 85 \quad c = -7.48) \tag{4-7a}$$

$$Y = 0.69F + 1.01F_C + 0.66P_F \quad (T^2 = 2.58 \quad \% = 79 \quad c = -3.92) \tag{4-8}$$

$$Y = 1.25F + 2.70F_C + 0.78P_F \quad (T^2 = 5.11 \quad \% = 86 \quad c = -7.80) \tag{4-8a}$$

The equations for women, 4-7a&8a, are of stronger significance than those for men, 4-7&8, which elevates all of the standardized coefficients, but with little change in their relative strengths. Figure 4-3 plots Equations 4-7&7a that were fit separately to men (solid line) and women (dashed line); these show the near identity of the two plotted lines to each other. Cases falling rightward and upward from the sloping lines are predicted to be YesA, and 89 of the 127 YesA men (70 %) are correctly classified in this way by Equation 4-7. Cases falling below and leftward are predicted to be NoA, and 60 of the 64 NoA men (94 %) are correctly classified in this way. For women, 78 % of YesA and 88 % of NoA subjects are correctly classified.

4.10. Comment

Figure 4-3 demonstrates a close association of atheroma jointly with foam cells and fibroplasia in the coronary intima. The association, as elaborated in Equations 4-7&7a, is independent of age, and very nearly independent of maleness; hence age and male effects on atheroma seem to be "explained" statistically. On the surface, Equations 4-8&8a appear to extract two aspects of intimal fibroplasia, F and F_C, which respectively contribute weakly and strongly to atheroma. This theme recurs in later chapters.

CHAPTER FIVE

THE SIZE OF THE SMC REALM ASSESSED WITH THE HELP OF SAMPLING THEORY

Abstract. Evidence from the use of sampling theory attests that some persons acquire enlarged realms of fibroplastic matrix around the intimal SMC's as a general property throughout the whole aortic sample space. Furthermore, later intrusion by atheroma seems to favor sites with the most fibroplasia per SMC while tending to avoid sites with many SMC's.

Some evidence indicates that intimal SMC's may be scattered over the lateral walls of the thoracic aorta rather like the scatter of pennies on a checker board [74]. This observation has far reaching biological consequences, especially in relation to the action of aging on the artery wall. The unique properties of the lateral thoracic aorta allow the testing of this question because, 1) sampling at distances greater than 1.3 cm assures minimal correlation between adjacent sites and 2) the lateral thoracic aorta seems to be a sufficiently level playing field (cf. Chapter Two).

Figure 5-1 illustrates the general principles using pools of representative NoA and YesA aortas (cf. Article 5.12). SMC numbers form frequency distributions that exhibit slanted bell shapes with upward trailing tails. In the YesA specimens, some sites are obscured by atheroma, and the distribution in Figure 5-1, right frame, concerns the remaining population of uncensored sites.

(Note: The plotted data are jagged because of restrictions upon the combining of cases into pools with only 190 cases in the available data set. For pooling, the means and the variances of SMC numbers must be similar or the pooled data will be bimodal or multimodal, and this keeps the numbers of specimens small in each pool.)

5.1. Modeling the distribution curves

The Poisson distribution would be especially convenient to describe the curves in Figure 5-1. The Poisson is the form taken by wholly random events. That convenient distribution is rejected, however, because the variance exceeds the mean in these pools of specimens, and in every aorta individually. The situation is reminiscent of pennies cast onto a grid. In a real tossing of physical pennies onto a checkerboard, the coins tend to pile up in aggregates. Although the aggregates seem randomly dispersed, the pennies are not, due to their clinging to each other in clusters. Because of this clustering, the distribution of pennies follows the negative binomial form [74].

The solid curves in Figure 5-1 represent the negative binomial distributions fit to the observed data (dashed lines). Chi Square tests fail to reject the fits, and therefore declare the fits to be consistent with the data. Table 5-1 reports the results of similar Chi Square tests applied to each of the 190 individual aortas in the series of

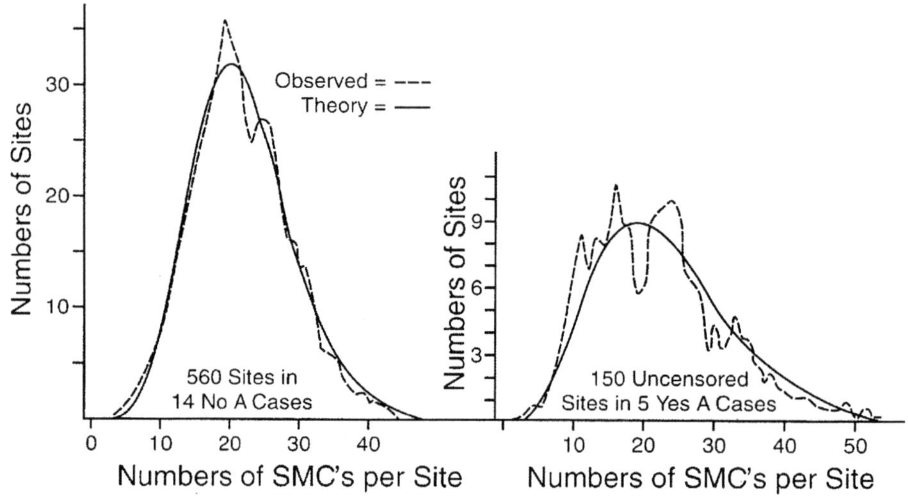

Figure 5-1. Negative binomial distribution curves are fit to data observed in two pools of aortas.

cases. The p-levels derived in this way form a pattern as expected for random events. The 14 aortas with $p<0.1$ could easily be attributed to chance alone, and the 14 departures from the theoretical expectation cannot be declared significant. It appears therefore, that exceptional aortas are rare or nonexistent.

The fitting of the theoretical curves to the YesA cases is remarkable. In these instances, the data sets are censored through obscuring of YesA sites by atheromas, which might be expected to distort the curve. The possibility arises that the censoring could be happening at random with respect to SMC numbers. This option is soon to be rejected.

5.2. Theory

The negative binomial arises when clusters are Poisson dispersed, and the numbers of individuals per cluster can be fairly described by the logarithmic distribution [2,19,39]. Jones and Mollison [24] used this device to describe the growth of bacteria in a culture dish. Whereas the colony counts per dish might follow a Poisson distribution with mean M_1, the numbers of bacteria per colony were described by a logarithmic distribution with mean M_2. Consequently the bacterial counts, X, followed the negative binomial with mean M_1M_2. That pattern inspires a conjecture on SMC scattering: In histologic sections, hypothetical clusters of SMC's offer no visible boundaries, and cells per

Table 5-1. Numbers of Chi Square tests comparing negative binomial forms of distribution to observed data on intimal SMC numbers by probability levels in190 aortas.

Probability of Chi Square	Numbers of tests*			
	YesA	NoA	Total	Expected
>0.9	6	10	16	19.0
0.8-0.9	6	13	19	19.0
0.7-0.8	7	14	21	19.0
0.6-0.7	6	16	22	19.0
0.5-0.6	4	14	18	19.0
0.4-0.5	8	17	25	19.0
0.3-0.4	4	20	24	19.0
0.2-0.3	2	13	15	19.0
0.1-0.2	5	11	16	19.0
<0.1	4	10	14	19.0

* Chi Square comparing the 10 pairs of total and expected values is 7.05; $0.6 > p > 0.5$

cluster is a quantity inaccessible to precise observation. Yet, the fitted negative binomial is accessible to perusal and its properties might allow us to work backward to deduce the inaccessible properties of the hypothetical clusters.

Taking the observed mean and variance of the fitted negative binomial in the left frame of Figure 5-1 to be C_M and C_V, we formerly found [74]

$$M_1 = C_M \log(C_V/C_M)/(C_V/C_M - 1) \qquad (5\text{-}1)$$

$$M_2 = C_M/M_1 \qquad (5\text{-}2)$$

The values of C_M and C_V yield $M_1 = 12.9$ and $M_2 = 1.6$ when entered into equations 5-1 and 5-2. For this pool of aortas we can deduce that an average cluster should contain 1.6 SMC's if the SMC's are adequately described by the logarithmic distribution. We further deduce that the average site should hold 12.9 clusters of SMC's.

5.3. Sizes of SMC clusters increase in YesA aortas

Equations 5-1 and 5-2 were applied to each of the 190 aortas, using the censored data sets for YesA cases. Table 5-2 presents the means of M_1 and M_2 with the NoA cases stratified by age. In NoA cases, M_1 reached a plateau from ages 20-69 years ($p = 0.14$

Table 5-2. Means of fibroplasia per SMC, F_C, and of hypothetical parameters for the logarithmic, M_2, compounded into the Poisson, M_1, to form the negative binomial fit to the distribution of SMC's over the lateral thoracic aorta in 52 YesA aortas and by age in 138 representative NoA aortas.

Age; years	M_1	M_2	F_C	N-cases
		--- NoA Aortas ---		
10-19	7.5	1.8	6.2	10
20-29	10.6	1.8	6.5	33
30-39	11.9	1.9	6.9	33
40-49	11.9	2.1	9.7	19
50-59	13.7	2.0	9.5	26
60-69	11.7	2.1	11.7	17
		--- YesA Aortas ---		
10-69	11.7	2.6	15.2	52

by ANOVA among ten-year age groups). YesA cases have M_1 like those in NoA cases. The shift of aging subjects into the YesA class is not selective, and should not bias the aging trend seen in the NoA groups. The values for M_2 fail to vary with age even before age 20 years ($p = 0.32$ by ANOVA).

YesA subjects have M_2 about 1/3 greater than NoA subjects. This could mean that M_2 increases in YesA cases with the onset of atheromas. On the other hand, perhaps the YesA subjects are removed from the NoA class selectively, and therefore leave a biased lot still in the NoA class. For this reason, the failure of M_2 to change with age in NoA subjects is of ambiguous meaning. Perhaps the bias disguises an increase in M_2 with age.

The total numbers of SMC's are reflected in the product of M_1 by M_2. This rises up to age 30 years and appears to stabilize after that age ($p = 0.37$ by ANOVA). The product rises modestly to 30.4 SMC's in the YesA cases from 24.7 SMC's in the NoA cases over age 30 years.

5.4. Avoidance by atheroma of large SMC clusters

In YesA cases the distribution curve of SMC numbers flattens near the mode, when compared with the NoA curve (Figure 5-1). The effect can be explained in two ways. 1) The broadening of the curve while retaining a constant mode might take place over time in a way that predisposes the aging aorta to atheroma. 2) The sites with many SMC's in the upper tail might fail to vanish with censorship by atheroma because of intrinsic resistance to the necrotizing process. Later evidence leads to a blending of these possibilities.

1) Some aging aortas are seen to rearrange the way that SMC's are scattered, thus reshaping their youthful tall, narrow curve toward a broadened, flattened shape. This could mean that fewer SMC's remain solitary while more of them are clustered in twos and threes and more. A way to explain this could be that enlarging of the realms by extruding bulky matrix materials might be chiefly a property of the solitary SMC's, with the clusters of two and three and more SMC's failing to perform this expansion as briskly. The clusters then become proportionately more numerous in each unit volume of space. The solitaries become diluted out of the field of view while the clusters are less prone to that effect. This mechanism fails to account for the modest increase in SMC numbers in the censorship-surviving sites in the YesA specimens.

2) The chief target for atheroma might be the sites where solitary SMC's occupy expanded volumes of space. This targeting would also tend to spare the sites with clustered SMC's. The censoring by atheroma could therefore preserve the upper tail of the curve while diminishing the middle and lower reaches. This kind of biased selection would elevate the mean of SMC numbers in the data set remaining after censorship. This mechanism fails to account for the broadening of the curve.

Explanations 1 and 2 are complementary to each other. One accounts for the broadening of the curve in YesA specimens while the other accounts for the elevated mean of SMC numbers. Both explanations call upon the same pattern of behavior by the SMC's. Those that exude excessive matrix materials become spread apart into solitaries, leaving the others still in clusters. Thus the solitaries are favored for obscuring by atheroma because of their excessive fibroplasia acting as lipid traps. Chapters Six and Seven will further pursue this topic.

5.5. The realm of the SMC

When the SMC's are scattered over the lateral aorta of a child during growth and development, each of them carries with it a certain bulk of collagenous matrix material. The imaginary boundaries of this material delineate a kind of "realm" for the SMC, which can be pictured as a brick shaped volume of space. If we could know the dimensions of the realms for each SMC, then we could explore the shapes of their distribution curves, as in Figure 5-1. Direct observation is prohibited, because the boundaries of the hypothetical "realms" are not generally visible in tissue sections. However, reasoning as before, we can use what we know so far to work backward and deduce the inaccessible properties of the hypothetical realms.

Table 5-3. Means and variances of SMC numbers averaged within pools of 52 YesA and 138 NoA aortas; how to calculate k.*

	NoA	YesA
SMC numbers		
Mean; C_M	21.6	27.9†
Variance; C_V	79	174
c-parameter	3.7	6.2
F_1; µm		
Mean; M_F	184	414
Variance; V_F	5701	55674
Ratio	31.0	134.5
Realm; k-parameter	8.4	21.7

* $c = C_V/C_M$ and Ratio = V_F/M_F and k = Ratio/c
† C_M of YesA cases exceeds that of NoA cases, perhaps because of biased selection when atheromas avoid the most cellular sites in an aorta.

The 190 aortas incorporated into Table 5-2 provide observations on SMC's to use for exploring the ratio of the variance, C_V, to the mean, C_M, $C_V/C_M = c$. Assuming that each SMC occupies a realm of constant size k, it follows that the numbers of realms should have variance and mean of similar ratio, $V_R/M_R \approx c$.

$$C_V/C_M = c \approx V_R/M_R = (V_F/k^2)/(M_F/k) \tag{5-3}$$

where V_F and M_F refer to the observable quantities, variance and mean of F_1. Hence

$$k \approx (V_F/M_F)/c \tag{5-4}$$

Table 5-3 gives two examples of how to calculate k, using variances and means averaged within the YesA and NoA pools of cases.

5.6. The sizes of realms estimated in two ways

Table 5-4 provides a variety of estimates for parameter c to use in equation 5-4. The c values obtained for each aorta through the leftmost side of equation 5-3 are averaged within intervals of F_1. Those average values range from 2.1 in the lowest F_1 group of NoA cases to 6.8 in the highest F_1 group of YesA cases. With these estimates in hand, we can proceed to calculate k through equation 5-4 for each case, and to average these. The quantity, k, is the thickness dimension of the hypothetical realm, which

Table 5-4. Means of SMC numbers (C_T), fibroplasia per SMC (F_C), and theoretical parameters of hypothetical realms in the lateral thoracic aorta by fibroplastic intimal thickness, F_I, in 190 representative aortas.*

F_I	C_T	c-param	F_C	Realm	N-cases
		--- NoA Cases ---			
<50	7.9	2.1	4.8	2.6	8
50-99	15.0	3.0	5.0	4.2	24
100-149	19.3	3.2	6.4	5.7	31
150-199	25.1	2.8	7.8	5.4	13
200-249	26.0	3.3	8.9	8.7	22
250-299	27.3	4.4	10.7	8.2	20
300+	25.6	4.2	13.6	9.6	20
		--- YesA Cases ---			
<300	28.8	4.0	8.3	9.1	9
300-349	26.7	5.6	11.6	20.6	5
350-399	24.3	5.0	15.1	25.3	15
400+	30.9	6.8	19.6	21.5	21

* The mean of SMC numbers is C_T in an aorta and is C_M in a pool of aortas.

corresponds to a related quantity, fibroplasia per SMC, F_C, as introduced in Chapter One. Both of these quantities refer to the space occupied by an SMC. However, F_C is taken from direct observations while k is a hypothetical entity derived indirectly from other sorts of data. Estimates of these two quantities are expected to bear serious imprecision because of measurement errors and statistical inefficiencies. The agreements therefore seem surprisingly good in the listings of Table 5-4, and in the scatter plot of individual cases in Figure 5-2.

The values for F_C in NoA aortas exceed those for k = realm in Table 5-4 by a factor ranging from 1.03 to 1.85, and averages 1.19. These values can be brought together by a small adjustment in the units used to measure $F_C = F_I/C_T$. The counting of SMC numbers, C_T, is done within an area that is arbitrarily set at 100 μm width (cf. rectangular regions in Chapter One, Figure 1-1A&B). By adjusting this to 119 μm width, the averages of F_C and k should tend toward equality. This result suggests 119 μm as an anatomic unit that is in some way "natural" to the system under discussion. This matter is considered further in Chapter Nine.

In Figure 5-2, the swarm of data points depart consistently from the line of $F_C = 1.19k$ only in the lower left portion of the plot. These subjects include some that

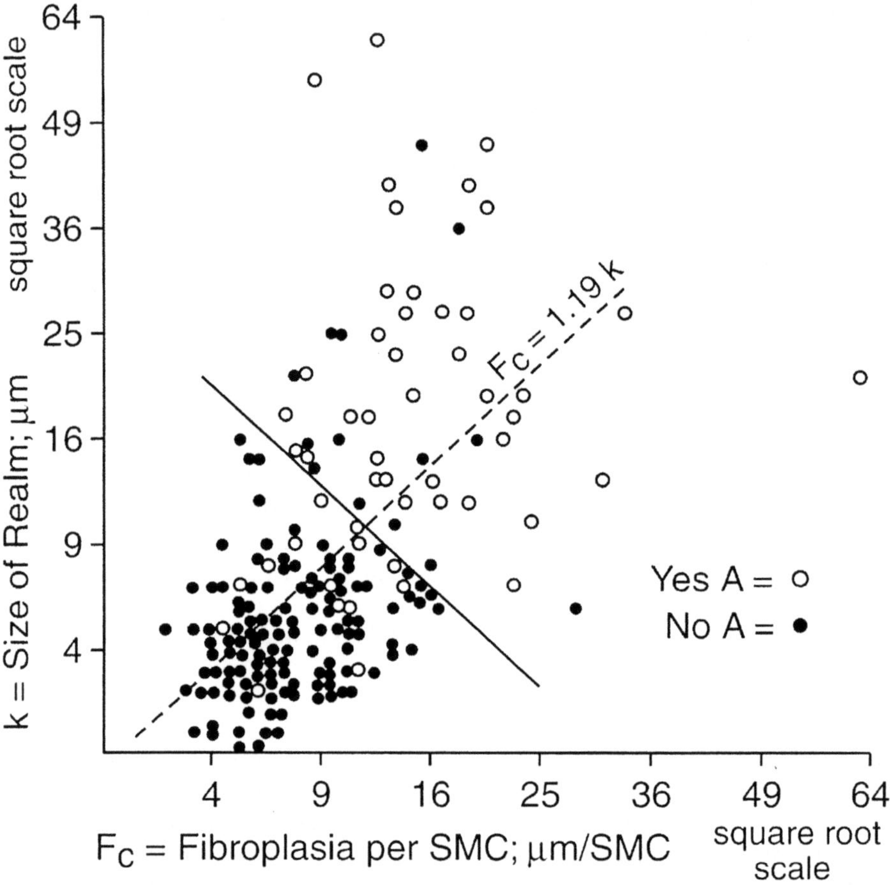

Figure 5-2. Dashed diagonal line represents F_C = 1.19k, and is not fit to the data points. Downward sloping solid line optimally separates dots from circles; it is the discriminant function (Equation 5-5).

haven't completed pubertal growth and maturation. The "natural" unit of width in these immature subjects appears to be near 185 µm rather than the mature value of 119 µm.

5.7. Influence of realm size on atheroma

In Chapter One, Figure 1-3 demonstrated a profound influence of F_C on the probability of finding atheroma in specimens of coronary artery. A similar outcome is now obtained for the quantity, k, representing the hypothetical realm obtained through equation 5-4 (Figure 5-2). The two variables, k and F_C, as measures of the same theoretical entity, space around the SMC, are expected to contain largely the same information. Combining these two sightings aimed at the same target helps to discriminate YesA from NoA aortas as shown in the scatter plot.

$$W_A = 2.32\sqrt{k} + 0.91\sqrt{F_C} \quad (T^2 = 3.18 \quad \% = 84 \quad c = -7.74) \tag{5-5}$$

The downward sloping solid line in Figure 5-2 represents equation 5-5, the discriminant function that optimally distinguishes YesA from NoA specimens when $W_A = 0$.

(Note: Symbols are explained in Article 1.7, Chapter One. Equation 5-5 and Figure 5-2 use square root transforms of k and F_C to correct for severe skewness in their distributions. Standard deviations of the square rooted variables are 1.92 $\sqrt{\mu m}$ and 0.79 $\sqrt{(\mu m/SMC)}$ respectively.)

The downward sloping solid line separating YesA from NoA symbols in the scatter plot centers near k = 10.7 μm (the point estimate found using a version of equation 5-5 with k alone). Thus, an increase in the size of the realm beyond a thickness of 10.7 μm imparts a high probability of finding an atheroma somewhere in the specimen. A comparable estimate of F_C = 9.3 μm was obtained for coronary arteries in Chapter One. The disagreement is 1.15 = 10.7/9.3, in keeping with the related disagreement of 1.19 discussed earlier.

5.8. How realms are dispersed

The observed values of F_1 for each aorta or for each pool of specimens can now be transformed into the hypothetical quantity, numbers of realms, through the relationship, N-realms = F_1/k, at each site in the specimen. Figure 5-3 presents some examples of the frequency distributions that N-realms describe. The specimens used here are taken from a different source than used above, thus broadening the generality of the conclusions (cf. Article 5.12). Estimates of c are drawn from Table 5-4, but the previous source of specimens does not otherwise enter the following discussion.

Figure 5-3 plots a pool of 10 NoA aortas chosen from the midrange of mean and variance of N-realms (upper frame). For this pool of specimens, the modal value of N-realms fell near 30, and k = 9.2 μm was estimated, because the mean of F_1 was 9.2x32.1 = 275 μm. The fitted negative binomial curve is not rejected by Chi Square test, and that failure of rejection is a necessary occurrence for the survival of our theory. The resulting curve is strikingly similar to Figure 5-1, left frame.

The pool of 10 YesA aortas is approached with caution, because it is a censored data set. Using only the 331 uncensored sites without atheroma the modal

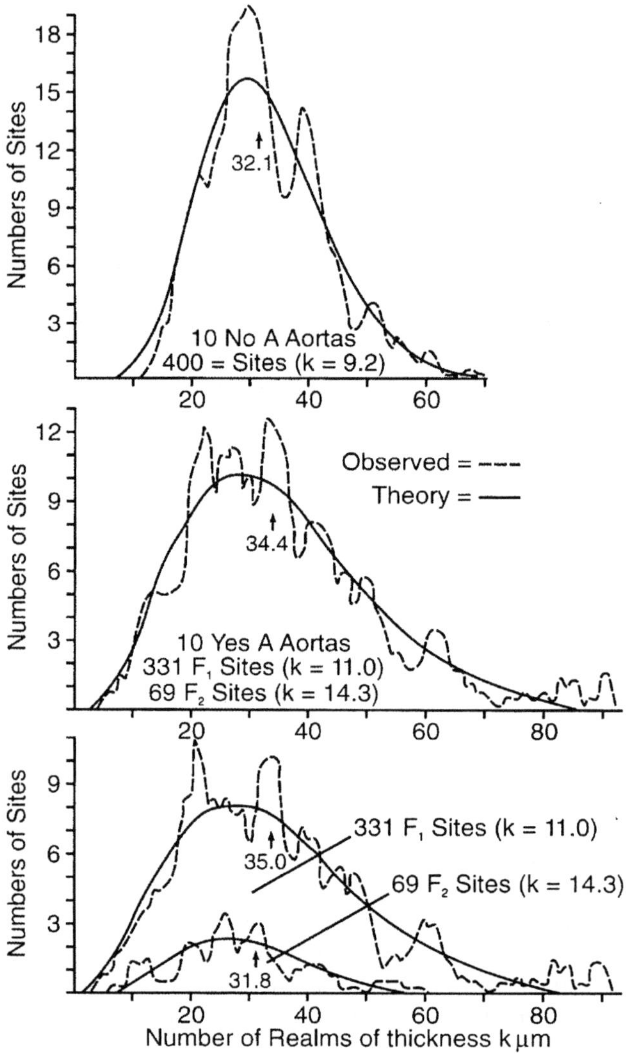

Figure 5-3. Hypothetical realms of SMC's are estimated to be of k µm thickness. A pool of 10 NoA aortas of similar characteristics yielded the frequency distribution in the top frame (dashed line) having a mean of 32.1 realms/site. The observed sites within the censored data set of 10 YesA aortas yielded the upper curve in the bottom frame (mean = 35.0 realms/site), while the estimated missing values in the censored data set yielded the lower curve (mean = 31.8 realms/site). The middle frame presents the combined curve obtained by adding the upper and lower curves of the bottom frame. Solid curves are the negative binomial fits to the data, and these are not rejected by Chi Square tests.

value of N-realms was again near 30, and k=11.0 μm per realm was found, because the mean of F_1 was 11.0x35.0 = 385 μm. The distribution of N-realms is plotted in Figure 5-4 (bottom frame, upper curve). The censored data set continues to obey the expected negative binomial form of distribution just as it did before in Figure 5-1, right frame. Again, the finding raises the possibility that the censoring happens at random with respect to N-realms. Here, also as before, further inquiry again tends to reject this interpretation.

5.9. Estimating missing values

The sites obscured by atheroma are not entirely without information. Although counting of SMC's is clearly impractical at atheromatous sites, the estimating of realms invites an indirect approach. As explained in Chapter Three, a quantity F_2 = cap plus base is observed and measured along with each atheroma. This quantity can be used to estimate realm size k through equation 5-4, and k = 14.3 μm is found for the pool of 69 atheromatous sites in these 10 YesA aortas. N-realms = F_2/k is then found just as for F_1. Figure 5-3 depicts the distribution of N-realms of F_2 (bottom frame, lower curve). The two curves in the bottom frame of Figure 5-3 look very much alike. Chi Square tests fail to reject the negative binomial fits. When the two curves in the bottom frame are added together to produce the middle frame of Figure 5-3, the reconstructed pattern with censored sites restored continues to fit the negative binomial form.

The patterns in Figure 5-3 generally tend to reiterate explanations 1 and 2 that arose in discussions of SMC numbers in Figure 5-1. The upper tail of the F_2 curve looks diminished compared with the long upper tail of the F_1 curve; the regions of aorta affected by atheromas tend to avoid the sites with the greatest numbers of realms. The mean/variance of the F_2 and F_1 curves are respectively 31.8/198.1 and 35.0/305.1 (bottom frame). Selective obscuring of the middle and lower reaches of the curve biases upward the mean of N-realms in the censored data set of F_1 sites while broadening the curve, just at it did earlier with SMC numbers. Estimates of hypothetical parameters M_1/M_2 yield 16.2/2.0 for the top frame, 9.9/3.5 for the middle frame, 9.8 /3.6 for the bottom frame upper curve, and 11.1/2.9 for the bottom frame lower curve. This increase in M_2 in YesA aortas implies a shift toward larger clusters of realms, just as it formerly did for the larger clusters of SMC's.

5.10. Goodness of fit

N-realms was calculated at each of 40 sites in 50 NoA aortas, and at an average of 30 uncensored sites in 33 YesA aortas chosen to have 25 to 35 such uncensored sites. Curve fittings like those in Figure 5-4 were carried out for each aorta, and the p-levels of the Chi Square goodness of fit tests tallied in Table 5-5. The pattern is in keeping with random expectation, implying that the 90 curves in this analysis are acceptably

Table 5-5. Numbers of Chi Square tests comparing negative binomial forms of distribution to observed data on hypothetical realms by probability levels in 83 aortas*.

Probability of Chi square	Numbers of tests			
	NoA	YesA	Total	Expected†
>0.9	5	4	9	8.3
0.8-0.9	5	3	8	8.3
0.7-0.8	4	5	9	8.3
0.6-0.7	4	6	10	8.3
0.5-0.6	5	2	7	8.3
0.4-0.5	7	1	8	8.3
0.3-0.4	4	2	6	8.3
0.2-0.3	4	0	4	8.3
0.1-0.2	5	3	8	8.3
<0.1	7	7	14	8.3

* For this analysis, k = size of a "realm" was tailored to each aorta so that k = (variance/mean)/c with c drawn from Table 5-4.
†Chi Square = 7.5, .7 > p > .6, for agreement of "total" with "expected".

well described by the negative binomial form. Exceptional aortas seem to be few or nonexistent.

5.11. Variation of k within aortas

A further complication arises from drawing together several diverse sources of information: Chapter Three reported a profound preference of atheroma for sites with the greatest fibroplastic thickenings, $F = F_1$ or F_2 as appropriate, within aortas as well as between aortas (cf. Tables 3-1 and 3-3). That result recurs here, because the mean of $F_2/F_1 \pm$ SEM is 1.24 ± 0.064, indicating that F_2 exceeds F_1 by an average of 24 % in the 33 YesA cases of this study. Figure 5-3 now informs us that the estimated missing values of N-realms in the pool of 10 representative aortas carry an augmented realm size, k=14.3 μm per SMC compared with k = 11.0 μm per SMC in the observed subset of censored data. This implies that atheroma selects for censorship the sites with the largest realms, where SMC's reside singly and widely separated by large bulks of fibroplasia per SMC. Figure 5-2 supports this conclusion in comparisons between aortas. The evidence just reviewed now supports this conclusion for comparisons from place to place within aortas. (The variation of F_C from place to place within each aorta is explored in Chapter Eight, where its impact upon the fitting of negative binomial curves is pursued.)

5.12. Sources of subjects

The two pools of aortas in Figure 5-1 were drawn from a series of 190 cases representing men and women aged 15-69 years in eight populations from various parts of the world [65]. The examples in the diagram were taken from the mid range of mean and variance of SMC numbers to assure that pooling of cases is appropriate. These restrictions on pooling criteria allowed 14 cases to enter the NoA pool, but only 5 cases into the YesA pool. In a sense, this series provides hypothesis formation.

The data set used for analyses in Figure 5-3 and Table 5-5 was taken from a series of 144 cases independent of that in the earlier series. This series of cases [57] was assembled in New Orleans from a forensic service, and represents all causes of death in men and women of ages 15-69 years. Of the 94 YesA specimens, 33 had 5 to 15 of the 40 measured positions "censored" by atheromas, and these are used in the table and figure. In a sense, this series provides hypothesis testing.

Chapter Six

Biased censoring of low SMC sites by atheroma in coronary artery

Abstract. Earlier data on aorta suggested that atheroma selectively censors arterial sites with the fewest SMC's, gaining this result from direct use of sampling theory in the level playing field of that organ. Here the same conclusion is obtained for the coronary artery, using an indirect approach adapted to the uneven playing field of this organ

SMC numbers, C_T, tend to vary in proportion to intimal thickness from place to place along the first eight centimeters of the right coronary artery. The variation is partly a systematic decline as the artery tapers from large to small along its length; but it is also partly random fluctuation. As the two variables, C_T and F (fibroplastic intimal thickness, cf. Table 1-1 in Chapter One) both decline, they remain in direct proportion to each other, but only in the arteries that lack atheromas, the NoA arteries. The proportionality is usually lost in arteries that harbor one or more sites of atheroma, the YesA arteries. Moreover, the severity of the loss increases as atheroma consumes ever more of the artery, as shown in the data of this chapter. The patterns raise the possibility that atheroma may preferentially select the fibrotically thickened sites that have the fewest SMC's, in confirmation of the same suggestion in the aortic findings of Chapter Five.

6.1. Photos of typical appearances

Frames “1" and “4" in Figure 6-1 depict the thinnest intimal sites observed along the eight centimeter samples of the representative NoA and YesA coronary arteries respectively. Frames “3" and “6" depict the thickest sites, while “2" and “5" depict examples of “average” in the sense of having measurements near the means of intimal thickness, F, and SMC numbers, C_T. The thinnest site in the NoA example (Frame “1") has $C_T = 13$ SMC's in the rectangular area, 100 μm wide with a height of F = 47 μm intimal thickness, giving $F_C = 47/14 = 3.6$ μm/SMC. The average (Frame “2") and thickest (Frame “3") sites have $F_C = 73/20 = 3.7$ μm/SMC and $F_C = 160/28 = 5.7$ μm/SMC respectively. The near constancy of these ratios illustrates the general finding of direct proportionality between F and C_T. In the YesA examples, the three ratios are respectively (Frames “4", “5", and “6") 80/13 = 6.1, 233/17 = 13.7, and 500/29 = 17.2 μm/SMC. Clearly, these ratios do not range around any constant value, confirming the violation of proportionality.

6.2. Plots of F by C_T converge at a region away from the origin of the graph

Four cases were chosen to illustrate, in Figure 6-2, some representative scatter plots of fibroplastic intimal thickness, F, by total SMC's, C_T. Each plotted point represents

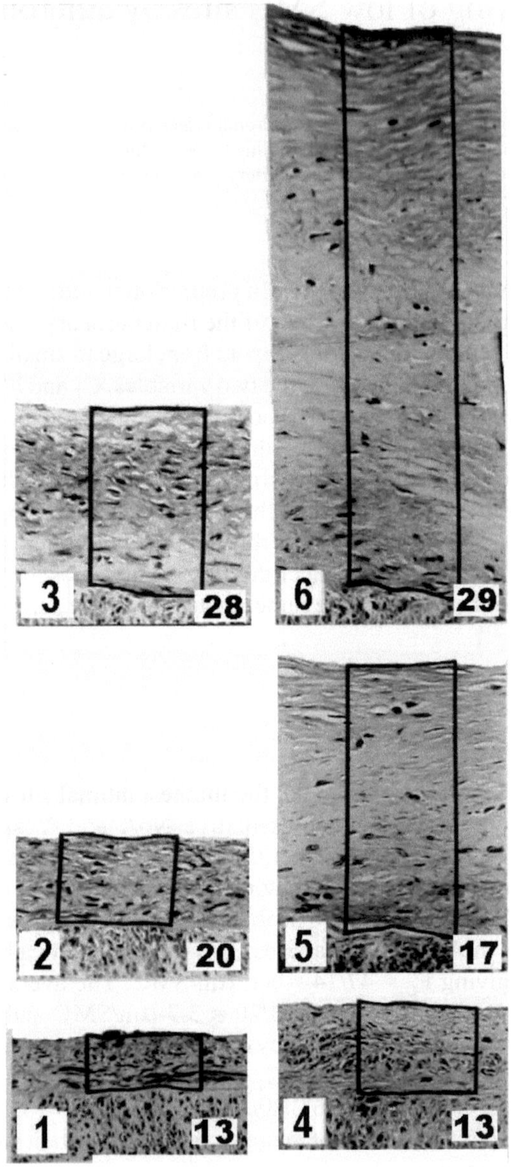

Figure 6-1. Typical appearances of a NoA (1, 2, & 3) and a YesA artery (4, 5, & 6) are shown; these are arteries "b" and "d" respectively in Figure 6-2. Rectangular areas are 100 μm wide, and subsume the full intimal thickness (internal elastica is placed with the media); numbers of nuclear profiles, counted in these areas under the x40 objective lens (lower right corners), estimate the intimal smooth muscle cell numbers and define the variable C_T. Areas compared side by side were chosen to have similar C_T. H&E.

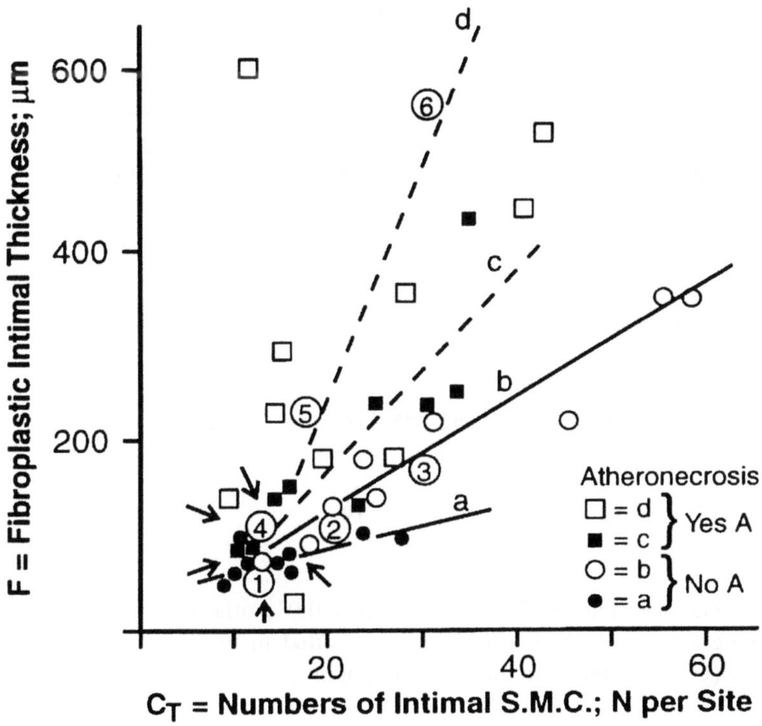

Figure 6-2. Four selected arteries, a-d, are shown. Plotted points represent the nine locations evaluated along the length of each artery, and the sloping lines represent the orthogonal regression equations that were derived from these data points. These arteries were chosen because their regression lines converge upon the area of the chart demarcated by arrows. Numbered positions, 1-3 in artery "b" and 4-6 in artery "d", correspond to the photographs in Figure 6-1.

one of the nine observed locations in arteries a, b, c, and d. Sloping lines represent the orthogonal regression equations fit to the four elliptic scatters of points. The four sets of data points used to construct Figure 6-2 were chosen to be "typical". They represent a full range of slopes, and their regression lines converge upon a commonly recurring part of the graph (demarcated by a ring of inward pointing arrows).

Each regression line shown here is fit through the principal axis of the elliptic swarm of data points by orthogonal regression. This is done to avoid problems of bias arising from regression to the mean that characterize lines fit by the usual least squares method (cf. Article 6.16).

Table 6-1. Means of slope and intercept coefficients in orthogonal regression of intimal fibroplasia (F) on total SMC's (C_T) in cases grouped by age and presence or absence of atheroma (YesA)†.

Age; years	Slope = S_O; µm per cell		Intercept = I_O µm		Number of Cases	
	NoA	YesA	NoA	YesA	NoA	YesA
15-29 (G_1)	4.6	6.6	15	14	51	5
30-44 (G_2)	5.9 **	14.9	7	**-127	48	32
45-60 (G_3)	5.5 **	16.0	51	**-110	21	35
61-88 (G_4)	15.5	13.1	-160	-19	5	13

** = YesA differs significantly from NoA at $p < 0.01$; otherwise $p > 0.05$. The equations having these average coefficients are plotted in Figures 6-3 & 6-4, and symbols (G_1) to (G_4) correspond to those in the graphs.

†By linear regression, $I_O = -14.3S_O + 106.1$ ($R^2 = 0.82$, $n = 8$); hence $C_T = 14.3$ and $F = 106.1$ at the region of convergence of the 8 pools of orthogonal regression lines having the average coefficients tabulated here; when plotted, this point falls in the area marked out by arrows in Figure 6-2.

6.3. Regression lines in NoA arteries

The two NoA coronary arteries plotted as solid lines in Figure 6-2, a and b, show regression lines projecting near to the graphic origin, in keeping with the general property of direct proportionality between F and C_T in NoA cases. Statistical tests of this property require pooling of cases to avoid certain problems with regression analysis (see Article 6.16). In such pools, examined hereafter, the convergence of these lines near the graphic origin is confirmed consistently for NoA specimens, with rare exceptional cases to be taken up later. The YesA specimens are the ones that account for the commonplace exceptions to the convergence near the origin, thus generating a phenomenon in need of explanation.

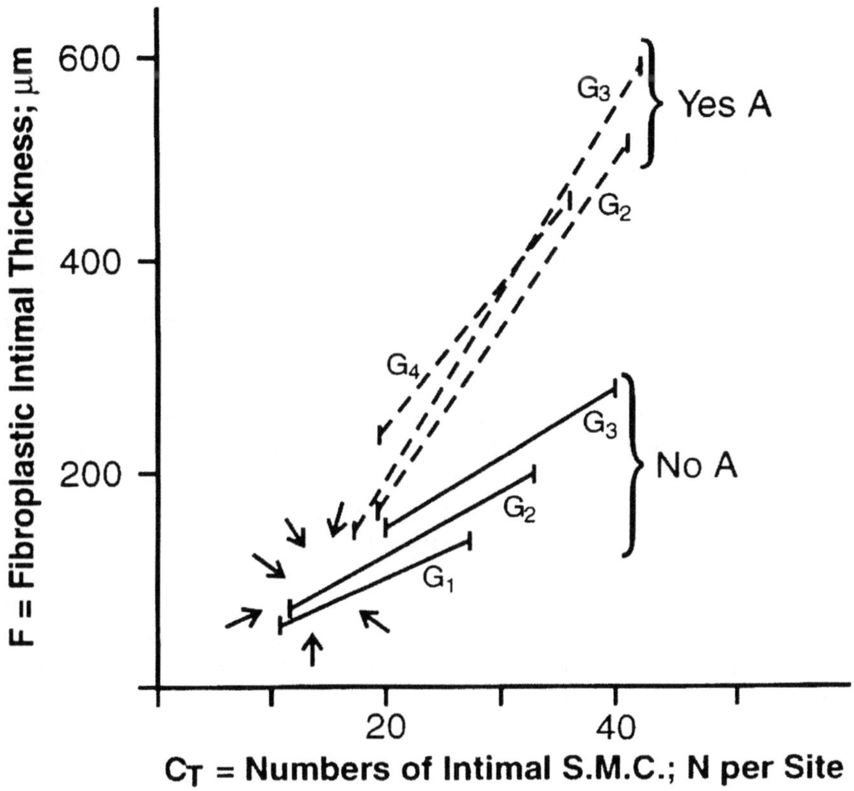

Figure 6-3. Graphed line segments, spanning the mean plus and minus one standard deviation, represent linear equations with the orthogonal regression coefficients given in Table 6-1, for YesA and NoA cases separately. The area demarcated by arrows is the same as in Figure 6-2. Symbols G_1, G_2, G_3, and G_4 represent age groups and correspond to those in Figure 6-4.

6.4. Regression lines in YesA arteries

The plots of two YesA coronary arteries as dashed lines in Figure 6-2, c and d, do not converge upon the graphic origin, but rather upon another location on the chart, in keeping with the loss of proportionality often seen in YesA cases. Confirmation of this behavior requires pooling of cases to overcome biases inherent in regression analysis, and this is done here in four different ways. *First way*: Age grouping is one useful way to construct pools (Table 6-1, Figure 6-3). *Second way*: This way is to cross tabulate cases according to C_T by F (Table 6-2 and Figure 6-4). *Third way*: This way is to cross tabulate cases according to C_T by P_A (Table 6-3 and Figure 6-5). A *fourth way* consists of combining all YesA and NoA cases into two pools to obtain a pair of simultaneous equations.

6.5. First way: age groups

Age groups G_1, G_2, and G_3 represent ages 15-29, 30-44, and 45-60 years in Figure 6-3.Regression slopes for the NoA subjects of these ages all project near to the graphic origin (G_4 is omitted because of containing few cases). Table 6-1 substantiates these conclusions, because, with one exception, the intercepts lie near zero in the averages of NoA arteries. The exception is formed by the pool of five NoA subjects over age 60 years, with orthogonal slope $S_o = 15.5$ μm/cell and orthogonal intercept $I_o = -160$ μm; these values resemble those of YesA subjects (which would happen if these five cases include some instances of false negative diagnosis).

Regression slopes for YesA cases are much steeper than the NoA slopes (G_1 is omitted from Figure 6-3 because of holding few cases). The G_2 and G_3 slopes fail to project toward the origin, but instead project toward the area demarcated by arrows. The G_4 pool with 13 members is ambiguous, with steep slope but intercept only modestly negative. Table 6-1 substantiates these findings.

Entries in Table 6-1 specify four equations for the pools of YesA cases. These can be solved simultaneously in pairs to estimate F and C_T where lines cross. The six estimates obtained this way average F = 171 μm and $C_T = 18.5$ cells/site; this point lies within the demarcated area in Figure 6-3.

6.6. Second way: groupings on F by C_T

Figure 6-4 illustrates another way of constructing data pools for averaging regression coefficients. NoA coronary arteries occur in five of the six regions defined in the graph; the other (Region 1) has only YesA coronary arteries. Regression coefficients in NoA cases show little variation between the five regions (Table 6-2, first and third columns); slopes are low and intercepts near zero in all five data pools.

Pools of YesA cases have variable outcomes (second and fourth columns of Table 6-2): Three groups, two of them with numerous members and one with only seven members, all show steep slopes and large negative intercepts (Regions 1, 2, & 4). Two other groups, with a total of only nine members, resemble NoA cases (Regions 3 & 5; YesA cases are absent from Region 6).

For general interest, regression lines representing age groups are included in Figure 6-4; for this purpose YesA and NoA cases are pooled within age groups. Sloping line segments that include the mean plus and minus one standard deviation represent the orthogonal regression equations for ages 15-29, 30-44, 45-59, and 60-88 years and

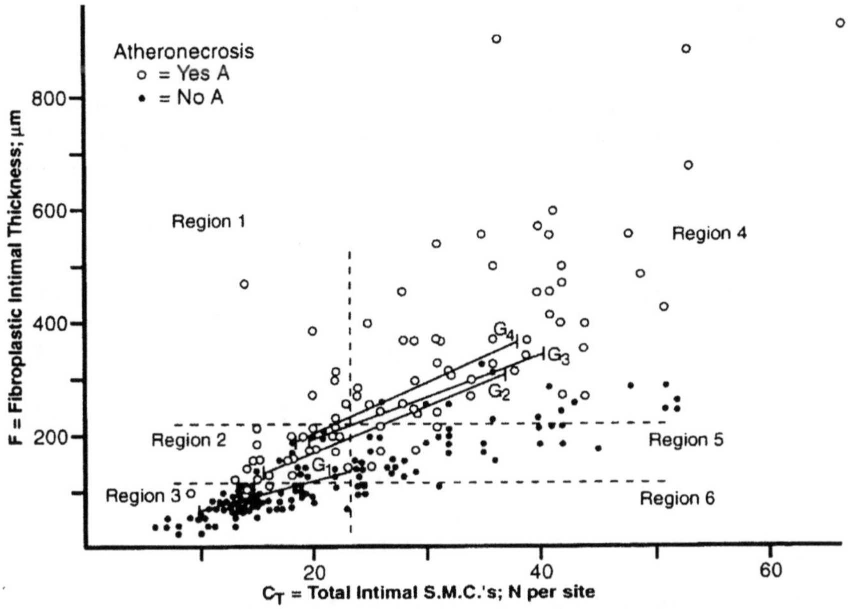

Figure 6-4. Mean values of cell numbers and intimal fibroplasia observed in 125 NoA (spots) and 85 YesA (circles) arteries form the scatter plot. Sloping lines represent regression line segments incorporating the mean plus and minus one standard deviation for the age groups 15-29, 30-44, 45-59, and 60-88 years (G_1, G_2, G_3, and G_4 respectively, YesA and NoA cases combined). Vertical dashed line divides the cases into two equal sized groups with more or less than 23 SMC's per site, and horizontal dashed lines define three equal sized groups using cutoff points of 119 and 224 µm of fibroplasia.

are marked G_1, G_2, G_3, and G_4 respectively. Circles, representing YesA cases, have a strong tendency to occupy the upper zones of the plot, because much fibroplasia per SMC, F/C_T, predisposes to the occurrence of atheroma, irrespective of cell total.)

The paired numbers in Table 6-2, slopes and intercepts, determine a set of five equations in the pools of YesA cases. Solving these simultaneously in pairs yields ten estimates of the points of intersection of regression lines. The averages, $F = 44$ µm and $C_T = 13.9$ cells/site, lie within the demarcated region of Figure 6-3.

6.7. The variable P_A

This variable is newly introduced here. It is measured in H&E stained paraffin sections by ink marking the portions of sample affected by atheroma. The affected portion,

Table 6-2. Means of slope and intercept coefficients in orthogonal regression of intimal fibroplasia (F) on total SMC's (C_T) in cases grouped by mean C_T and F according to regions in Figure 6-4†.

Region	Group C_T	Group F	Slope = S_O; µm per cell NoA	Slope = S_O; µm per cell YesA	Intercept = I_O µm NoA	Intercept = I_O µm YesA	Number of Cases NoA	Number of Cases YesA
1	low	high	.--	20.4	---	-115	---	7
2	low	mid	6.8	**17.7	26	**-143	13	21
3	low	low	5.1	6.7	10	18	61	4
4	high	high	8.9	*13.8	-36	-89	15	48
5	high	mid	5.1	7.4	30	-13	31	5
6	high	low	4.3	.--	-1	.--	5	.--

** YesA differs significantly from NoA at $p < 0.01$; * = $p < 0.05$

† By linear regression, $I_O = -10.4S_O + 68.6$ ($R^2 = 0.89$, n=10); hence $C_T = 10.4$ and F = 68.6 at the region of convergence of the 10 pools of orthogonal regression lines having the average coefficients tabulated here; this falls in the area marked by arrows in Figure 6-2.

expressed as a percentage of the specimen, is the meaning of P_A. As P_A increases, the censoring action of atheroma obscures ever increasing amounts of tissue sample.

6.8. Third way: groupings on C_T by P_A

In Figure 6-5, dashed lines divide the graph into six regions, which are examined in Table 6-3. The two NoA pools with $P_A = 0$ (Regions 1 and 2), have shallow slopes and near zero intercepts, as usual, in the tabulations. Pools of YesA cases again have variable outcomes. Three of four YesA groups have steep slopes and large negative intercepts, in the familiar YesA pattern (Regions 4-6); while the fourth pool is more nearly like the NoA pattern (Region 3).

The four equations for YesA pools in Table 6-3 yield six estimates of intersections between pairs of regression lines. The averages, F = 81 µm and $C_T = 12.7$ cells/site, lie within the demarcated region of Figure 6-3.

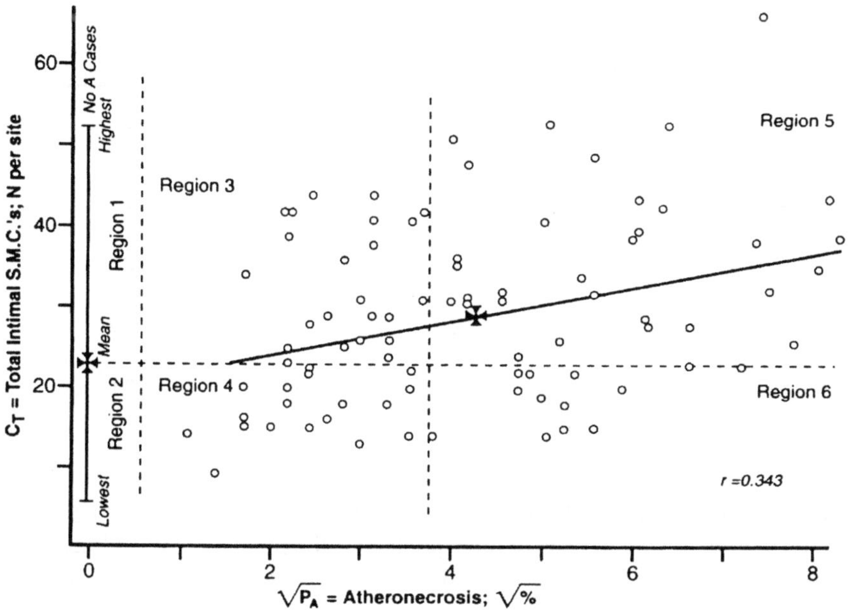

Figure 6-5. Mean values of cell total and atheroma (% of specimen) observed in the 85 YesA cases form the scatter plot; sloping line is the fitted least squares regression line which has r = 0.343. The NoA cases all have P_A = 0; their mean and range for C_T are depicted at zero on the abscissa. Horizontal dashed line marks the same cutoff point of 23 cells/site that was used for a vertical line in Figure 6-4. The vertical dashed line on the left separates YesA from NoA cases, and that on the right divides the YesA cases into two groups of equal size at the cutoff point of $\sqrt{P_A}$ = √14%.. Maltese crosses locate means of NoA and YesA pools of cases.

6.9. Fourth way: convergences of all regression lines

Each specimen in the collection of 210 coronary arteries is described by an equation of the form Y = bX + a, where Y = F, the fibroplastic intimal thickness at one of the nine observed locations in the artery, and X = C_T, the associated total number of SMC's. The empirical parameters, a and b, are determined by orthogonal regression over the nine data points in each case, so that b is renamed S_o and a is renamed I_O. S_o was determined to have an average of 5.7 μm per cell in the 125 NoA arteries, and I_O was determined to average 11 μm (data not shown). The respective averages in the 85 YesA arteries are 14.6 μm per cell and -95 μm. Substituting these two sets of values produces two simultaneous equations to use for estimating X and Y, yielding the simultaneous solution of C_T = 9.4 cells/site and F = 64 μm. These values define the

Table 6-3. Means of slope and intercept coefficients in orthogonal regression of intimal fibroplasia (F) on total SMC's (C_T) in cases grouped by mean C_T and extent of atheroma (P_A) according to regions in Figure 6-5†.

Region	Group C_T	Group P_A	Slope = S_O; µm per cell	Intercept = I_O µm	Number of Cases
1	high	zero	6.2	7	51
2	low	zero	5.4	13	74
3	high	low	9.3	-8	22
4	low	low	14.1	-100	19
5	high	high	16.0	-135	31
6	low	high	20.9	-141	13

† By linear regression, $I_O = -11.5S_O + 77.2$ ($R^2 = 0.92$, n=6); hence $C_T = 11.5$ and F = 77.2 at the region of convergence of the 6 pools of orthogonal regression lines having the average coefficients tabulated here; this falls in the area marked out by arrows in Figure 6-2.

point of intersection of the two lines. The point falls within the demarcated area in Figure 6-2.

6.10. Estimating cell numbers at censored sites

Figure 6-5 discloses a positive correlation of SMC numbers, C_T, with the extent of atheroma, $\sqrt{P_A}$, in YesA cases ($r = 0.343$). This is a recurring phenomenon that calls for explanation. One possible explanation explored here is that atheroma may selectively censor arterial sites with the fewest SMC's. The hypothesis cannot be rejected by the data after the quantitative testing that now follows:

The approach used here is to define certain invisible censored quantities and then to devise indirect approaches for estimating their expected magnitudes. The theory must be rejected if estimated quantities are unreasonable.

To formalize, let C represent cell numbers in rectangular areas such as those in Figure 6-1. C_M denotes SMC's that formerly occupied the obscured sites that are *missing* from the observed sample due to censorship. C_T denotes the SMC's in the observed part of the specimen, and C_B those in the whole artery *before* censorship. Also let p be the proportion of specimen obscured by atheroma, i.e. $p = P_A/100$. Hence, $C_B = pC_M + (1 - p)C_T$. In a series of n arteries, C_T and p can be measured in each of them, and from the n data pairs, C_B and C_M can be calculated as a and b coefficients in a regression equation, $Y = bX + a$, where $Y = (1 - p)C_T$, $X = p$, $C_M = -b$, and $C_B = a$.

The estimates of C_B and C_M obtained in this way are *averages* to be applied alike to all arteries. This generalization assumes that C_T correlates with P_A solely because of the biased censorship that atheroma imposes. Therefore, the adjustment for the effects of censorship should leave C_B uncorrelated with P_A.

The equation calculated from the 85 YesA cases of Figure 6-5 is

$$Y = -15.4X + 25.8 \ (R^2 = 0.086) \quad (6\text{-}1)$$

This gives point estimates of 15.4 cells/site ($C_M = -b$) as the *missing* cellularity at locations that were censored by atheromas, and 25.8 SMC's as the average cell total ($a = C_B$) presumed to pervade throughout the artery *before* the intrusion of necrosis. C_B, the hypothetically uncensored cell total calculated for each artery, does not correlate with p ($r = 0.04$), as predicted. This finding confirms that reasonable values of C_M and C_B can account in full for the correlation in Figure 6-5. Hence, the data do not contradict the theory.

6.11. Preparing for the estimation of fibroplasia at censored sites

The average fibroplastic intimal thickness associated with censored sites, say F_M, is missing information. Estimating the missing F_M, however, is not as straight forward as for C_M. Regarding C_M, this quantity changes little with age and is uncorrelated with atheroma after adjustment for missing information; these properties allow us to treat C_M as practically invariant in the averages of pooled cases. Regarding F_M, this quantity increases rapidly with age and with atheroma; its correlation with P_A is not ascribed to sampling bias. In the following development, we are unable to find point estimates for F_M, but only its upper and lower limits.

We now need to define two additional terms to prepare for approaching the estimation of F_M. These two terms begin by accepting the prior result $C_T = 15.4$ SMC's for the average of censored sites. In Figure 6-2, each regression line intersects the vertical line erected at $C_T = 15.4$ SMC's, and we shall name this intersection F_O. This quantity is calculated from each of the regression lines fit to the specimens:

$$F_O = 15.4S_O + I_O \quad (6\text{-}2)$$

Table 6-4. Statistical estimates of the upper limit for F_M, information missing because of censorship of arterial sites by atheroma, according to the percentage of specimen affected by atheroma.

$\sqrt{P_A}$; √% specimen with atheroma	Determined at C_T = 15.4 SMC: Observed by Regression F_O:µm	Estimated by statistical model F_S:µm	F_M:µm	Adjusted †
<1.5 (4)*	128	138	598	0.95
1.6-2.1 (5)	123	130	340	0.94
2.2-2.9 (21)	112	140	606	0.85
3.0-3.7 (18)	145	141	80	1.05
3.8-5.0 (21)	132	187	423	0.77
5.1-6.5 (21)	102	194	354	0.68
>6.5 (15)	166	190	238	0.94
ANOVA F	0.41	3.56	1.17	1.39
p	0.86	0.00	0.33	0.23

* Number of cases in parentheses
† Adjusted = $(p(F_S - F_O) + F/C_T))/(F/C_T)$

where S_O and I_O are the regression slope and intercept respectively for each of the 85 YesA cases.

By reading the graph of Figure 6-2, we can see that shifting the YesA lines to force them through the graphic origin requires rotating the lower end upward from the region ringed by arrows. Each rotated line intersects the vertical at C_T = 15.4 SMC's, and we shall name this intersection F_S. We can see that $F_S/15.4 > F/C_T$, i.e. the ratio of the shifted values on the left is larger than the ratio of the observed quantities on the right. We shall later find, however, that the final impact of the discrepancy is small and can be treated as negligible, allowing us to write.

$$F_S \approx 15.4F/C_T \tag{6-3}$$

Because Equation 6-3 contains a small inaccuracy, we must proceed from here with caution.

Table 6-4 reports mean values of F_O and F_S within strata of increasing extent of atheroma, P_A. F_O fluctuates strictly at random across P_A strata (not significant in the ANOVA). This outcome reiterates the convergence of regression lines within the demarcated regions in Figures 6-2&3. F_S increases significantly with P_A (strongly significant in the ANOVA). The adjustment upward needed to force each group of lines through the origin increases with the extent of specimen censored by atheroma. This result is the expected outcome if the trend is caused by the intrusion of atheroma.

6.12. Making the estimation of F_M

We are now in a position to attempt an estimation of the missing information, F_M, using

$$F_S = qF_M + (1 - q)F_O \qquad (6\text{-}4)$$

where q measures the censored proportion of sites in the region of $C_T = 15.4$ SMC's. If a value can be substituted for q, then F_M remains the only unknown so that it can be found. We began with the assumption that $q > p$, which is to say that the greatest proportion of atheromas should favor sites that originally held few SMC's. To remain consistent with the initial assumption, we can safely employ p as a lower limit of q. Seeing that F_M is an inverse function of q, we can find an upper limit for F_M by setting $q = p$.

Table 6-4 reports mean values for the upper-limit estimates of F_M. Six of the seven estimates are all within the range of 238 to 606 μm of intimal thickness (cf. circles in Figure 6-4 generally occupy this range). This range corresponds to 15.4 to 39.4 μm/SMC, which fits well into Table 5-4 in Chapter Five. The seventh estimate is $F_M = 80$ μm which is less than the observed value, F_O. The ANOVA, however, declares insignificant the variation among the seven strata, and the single disparity cannot be called significant. Taking the consensus of these estimated values as upper limits, and using F_S, perhaps, as lower limits, therefore identifies the general range of magnitudes required by the missing quantities that are needed to preserve our theory: 160 μm to 291 μm, the means of the two quantities F_S and F_M in Table 6-4. The actual value should fall within this range. Since the magnitudes are reasonable, the theory cannot be rejected by this line of reasoning.

6.13. A final check on consistency of reasoning

If the sites obscured by atheroma were those that previously held the greatest amounts of fibroplasia per SMC, then we should anticipate a biasing downward when calculating the ratio F/C_T solely from observed arterial sites. The magnitude of this downward bias is open to assessment by making use of the foregoing deductions. This is done through

$$(F/C_T)/(q(F_M - F_O)/15.4 + F/C_T) \qquad (6\text{-}5)$$

where the observed fibroplasia per SMC, F/C_T, is augmented in the denominator by the quantity presumed to be lost to censorship, $q(F_M - F_O)/15.4$, in the atheromatous proportion of specimen, q. This route again offers us an upper limit rather than a point estimate, because it assigns *all* atheromas to the region of $C_T = 15.4$ SMC's. The average magnitude of the downward bias of F/C_T is 0.88 and the variation among strata is insignificant in Table 6-4.

These results serve to assess the importance of the various assumptions that were made in the development of Equations 6-3&4. The accumulated maximum possible impact of these inaccuracies upon the end result is small (i.e. -12 %). Moreover, we fail to see a growing importance of these discrepancies as P_A increases, a trend that is expected if the effects had been conspicuous.

6.14. An alternative hypothesis

Suppose that sites with few SMC's are resistant to fibroplasia, rather than prone to atheroma. Data points might be missing in Figures 6-2&3, in the region above the rings of arrows, not because they were selectively removed but because they never formed. By this hypothesis, the estimated missing values for C_T would be telling us which sites tended to resist fibroplasia, and F_M telling us how much fibroplasia was resisted in this manner. The correlation of C_T with P_A might then be viewed as a real increase in SMC numbers at fibroplastic sites, i.e. that sites of fibroplasia are simultaneously affected by collagenous matrix expansion together with modest increases in SMC numbers.

This hypothesis fails to explain the peculiar parallel courses taken by lines G_1, G_2, G_3, and G_4 in Figure 6-4. As fibroplasia and atheroma worsen with age, the correlation of C_T with F remains unaffected. Other subtle discrepancies can also be found in the details of this chapter. The hypothesis also encounters further difficulties in the aortic data to be considered next, in Chapter Seven.

6.15. Sources of subjects

The coronary arteries examined here are among those described in Chapter One, Article 1.9. These were an early series, subsequently expanded by further case collection after completion of the analyses given in this chapter. Men and women are retained in a common pool because the "maleness" risk factor is taken to be subsumed in the general theory (cf. Chapter One, Figure 1-3 and Equation 1-3).

6.16. Correlation of regression coefficients of slope and intercept

An objective of this study is to explore the behavior of regression intercept terms. A problem arises, however, because slope and intercept coefficients have a mathematically obligatory inverse correlation with each other. Random errors that steepen the slope also simultaneously lower the intercept, and vice versa. Therefore, the search for a biological correlation between the coefficients requires *grouping* of cases, so that random fluctuations can cancel each other to nullify the spurious correlation. Pooling of data should minimize random effects, because the averages of random error will tend toward zero with increasing sample size. (The use of orthogonal equations also helps, because wide or narrow data ellipses have the same principal axis, allowing appropriate pooling of their equations. Moreover, the ambiguity of "two regression equations" is avoided, so that it is not necessary to deal with different intercepts resulting from regression of y on x versus x on y. Furthermore, the least squares line will have a nonzero intercept in a scatter plot whose principal axis intersects the origin.)

6.17. Magnitude of bias upon SMC numbers by atheroma censorship

In the present study, observed mean SMC numbers, C_T, were greater in YesA subjects than those of age matched NoA subjects (although not significantly so in the ANOVA, data not tabulated). After adjusting the cell totals for the biasing effect of atheroma, adjusted values have an estimated average of 25.8 cells/site in the overall pool of YesA arteries, compared with 24.8 cells/site in the pool of NoA cases at similar ages.

Chapter Six Appendix Table. Descriptive characteristics of selected variables, including correlation coefficients, raw (upper triangle) and age adjusted (lower triangle) in 210 specimens of coronary artery, men and women age 15-79 years.†

					Correlation coefficients					
Variable	Symbol	Units	Mean	SD	S_o	I_o	F	$\sqrt{P_F}$	C_T	$\sqrt{P_A}$
Age	A	Years	39.9	14.7	.37**	-.18**	.45**	.17*	.31	.41
Slope	S_O	µm/cell	9.3	9.2	.---	-.88**	.44**	.26**	.10	.49**
Intercept	I_O	µm	-32.1	191	-.89**	.---	-.25**	-.17*	-.05	-.28**
Intima (fibroplasia)	F	µm	209	55	.33**	-.19**	.---	.40**	.74**	.77**
Foam cells	$\sqrt{P_F}$	√%	3.1	2.3	.21**	-.14*	.37**	.---	.25**	.47**
Cell total	C_T	N/(100 µm)²	25.3	11.9	-.03	.01	.71**	.21**	.---	.36**
Atheroma	$\sqrt{P_A}$	√%	1.7	2.4	.40**	-.22**	.64**	.45**	.26**	.---

** significantly different from zero with $p < 0.01$; * = $p < 0.05$.
†P_F and P_A are measured in square root units to reduce the skewness of their distributions.

CHAPTER SEVEN

BIASED CENSORING OF LOW SMC SITES BY ATHEROMA IN THORACIC AORTA

Abstract. Atheroma tends to intrude into arterial sites that were previously prepared by aging effects. The dominating effect appears to be the expansion of the fibroplastic realms around the SMC's. Yet, some evidence seems to raise the possibility that, all else being equal, the fibroplastic sites with the fewest SMC's seem especially vulnerable.

This chapter uses data from the lateral thoracic aorta to develop the same lines of reasoning applied to coronary arteries in Chapter Six. This is done for three principal reasons: (1) to test the reproducibility of the findings in an independent study, (2) to compare separate arterial segments to determine what aspects of the data sets are alike or different in the varied parts of the arterial tree, and (3) to explore the importance of having the level playing field offered by the homogeneous set of arterial sites in the aorta.

The results found here for the aorta are amazingly similar to those in the coronaries. The intimal behavior in the large elastic artery is almost indistinguishable from that in the much smaller muscular artery. The simple principles uncovered in Chapter Six may therefore be of widespread generality, having been reproduced faithfully in two out of two studies under varied circumstances.

7.1. The broadened scatter of SMC's in YesA aortas

Two phenomena in the coronary arteries converged upon the conclusion that atheroma selectively censors the sites with few intimal SMC's. These phenomena were (1) the convergence of regression lines in a region of the graph encircled by arrows in Figure 6-2 and (2) the numbers of SMC's increased in YesA arteries and continued increasing with expansion of atheromas over greater amounts of specimen. A third phenomenon could not be examined in the coronary specimens, although it is accessible in the aorta: (3) The frequency distribution of SMC's over the arterial sites is broadened and flattened in YesA aortas, as shown in Chapter Five. The almost perfect homogeneity of sites along the lateral thoracic aorta permits the use of sampling theory to explore this third phenomenon.

Figure 7-1 diagrams the relationship of mean SMC numbers, C_T, to its standard deviation, S_C. The sloping line that optimally separates circles from dots is the discriminant function derived for distinguishing YesA from NoA aortas when W = 0.

$$W = 1.9S_C - 0.84C_T \quad (T^2 = 1.35 \quad \% = 74 \quad c = -2.52) \qquad (7\text{-}1)$$

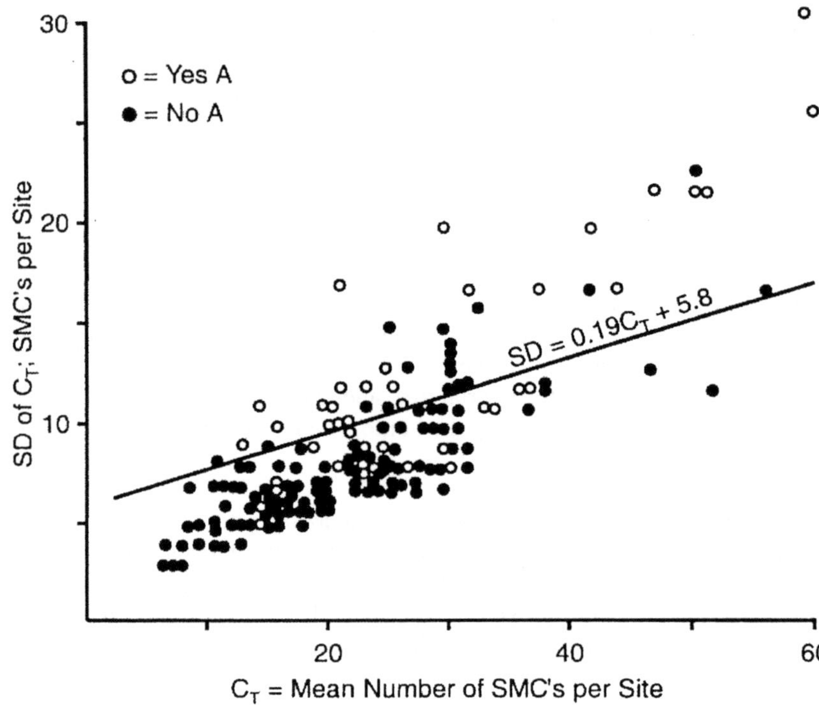

Figure 7-1. The standard deviation of SMC numbers is plotted against the mean, C_T, in 124 NoA and 48 YesA aortas. The sloping line represents the discriminant function that optimally distinguishes circles from dots; Equation 7-1 is the standardized form of this function.

(These are standardized discriminant coefficients, as explained in Article 1.7 in Chapter One; standard deviations are 4.39 for S_C and 10.29 for C_T). S_C averages respectively 14.5 and 8.8 SMC's per site in YesA and NoA aortas, a significant difference. C_T averages 30.8 and 23.0 SMC's per site in YesA and NoA aortas, also a significant difference. Figure 5-1 in Chapter Five illustrates the broadening and flattening of the C_T distribution curves of YesA specimens. Figure 7-1 acts to summarize the widening of the distribution curves as a vertical displacement on this graph.

(Technical note: Both of these variables contribute jointly to distinguishing the two types of specimen from each other, but the influence of C_T reverses in the presence of S_C. This outcome is apt to be a reflection of multicolinearity, seeing that r = 0.83 is the correlation coefficient linking S_C and C_T. Circles displaced upward from

Table 7-1. Means of orthogonal slopes and intercepts relating F to C_T in 172 aortas of subjects aged 15 to 69 years according to mean C_T and its standard deviation.†

Relation to Figure 7-1 discrim. line	Type of Aorta	S_O = Slope µm/SMC	I_O = Intercept SMC/Site	Number of cases
Below	NoA	8.4	3	100
	YesA	31.5	-299	23
Above	NoA	6.3	61	24
	YesA	25.8	-212	25
ANOVA		F (p)	F (p)	
Down or Up		2.2 (0.14)	2.8 (0.10)	
Yes or No		65.7 (0.00)	43.5 (0.00)	
Interact		0.5 (0.49)	0.1 (0.74)	

† By linear regression, $I_O = -13.6S_O + 133$ ($R^2 = 0.99$, n=4); hence $C_T = 13.6$ and F = 133 at the region of convergence of the four orthogonal regression lines having the average coefficients tabulated here; this falls in the area marked by arrows in Figure 6-2.

the slanted line in Figure 7-1 are also, *ipso facto*, displaced leftward from it. This aspect of Equation 7-1 implies that broadening of the distribution curves by elevating the ratio S_C/C_T clearly distinguishes YesA from NoA specimens as a class, but that shifting of the curve rightward by increasing mean C_T is of trivial consequence for the distinction.)

7.2. Regression lines in aortas with broadened SMC distribution curves

Regression lines are drawn here after the manner of those in Chapter Six, Figure 6-2. F and C_T are measured at each of 40 sites in the lateral walls of the NoA aortas and at N uncensored sites in YesA aortas. Orthogonal regression slopes and intercepts are calculated for each aorta, and these are averaged within pools of specimens constructed various ways. We shall later pursue the four ways of pooling cases explained in Chapter Six. But let us begin with Figure 7-1, which offers a way that is unavailable for coronaries.

Table 7-1 presents the averaged regression coefficients for the four pools of cases: YesA or NoA, above or below the line in Figure 7-1. As we previously saw for the NoA coronary arteries, the regression lines for the NoA aortas here again pass near the graphic origin of the plots. This finding is evidenced by the values for the intercept coefficients falling near zero, especially in the numerous NoA cases below the line. The regression lines for the YesA aortas also reproduce the pattern seen in YesA coronaries, averaging steep slopes with large negative intercepts.

The slopes and intercepts in Table 7-1 relate to equations of the form,

$$F = S_O C_T + I_O \quad (7\text{-}2)$$

Four versions of this equation are defined by the pairs of tabulated values for S_O and I_O. These four versions can be examined by linear regression to find F and C_T in the rearranged equation $I_O = -C_T S_O + F$, treating F and C_T as regression coefficients. These values for F and C_T are those that converge most closely in all settings, thus marking the region of convergence by the four equations of the form for Equation 7-2. They are found to be $C_T = 13.6$ SMC's per site and F = 133 μm (cf., footnote of Table 7-1). This falls, as for the coronaries, in the region of Figure 6-2 demarcated by arrows.

7.3. Region of convergence by regression lines

Chapter Six introduced four ways to find the convergence region where plots of Equation 7-2 in various data pools most often cross each other. *First way:* Age grouping. *Second way:* Cross tabulate cases according to C_T by F. *Third way:* Cross tabulate cases according to C_T by P_A. A *fourth way* consists of combining all YesA cases and NoA cases into two pools whose regression equations can be solved as a pair of simultaneous equations. These four ways, previously explored with coronary data, are reexplored here with the aortic data set.

7.4. First way: age groups

Ten-year age groups are used here in place of the15-year ranges employed for the coronaries. This is done to preserve consistency with Chapter Five. The averaged regression lines for the NoA subjects pass near the graphic origin in all of these age groups (Table 7-2, intercept terms, I_O, average near zero). Regression slopes for YesA cases are much steeper than the NoA slopes in all but the youngest age group, which holds four YesA cases, while intercept terms hold large negative values.

The ten versions of Equation 7-2 listed in Table 7-2 give estimates of C_T = 13.1 SMC's per site and F = 111 μm at the point of closest convergence (cf. footnote of Table 7-2).

Table 7-2. Means of orthogonal slopes and intercepts relating F to C_T in 172 aortas according to 10-year age groups.†

Age; years	S_O = Slope µm/SMC		I_O = Intercept SMC/Site		Number of cases	
	NoA	YesA	NoA	YesA	NoA	YesA
15-29	4.2	8.0	41	21	40	4
30-39	8.3	16.8	-11	* -247	31	3
40-49	11.0	20.9	-3	-188	21	4
50-59	10.2	**27.6	14	**-237	21	13
60-69	11.2	**35.2	26	**-321	11	24

ANOVA	F	(p)	F	(p)
Age	4.3	(0.00)	1.5	(0.22)
YesA	18.1	(0.00)	16.0	(0.00)
Interactions	1.7	(0.14)	1.2	(0.31)

† By linear regression, $I_O = -13.1S_O + 111$ ($R^2 = 0.84$, n=10); hence $C_T = 13.1$ and F = 111 at the region of convergence of the 10 pools of orthogonal regression lines having the average coefficients tabulated here; this falls in the area marked by arrows in Figure 6-2.

**=p<0.01 comparing YesA with NoA, *=p<0.05.

7.5. Second way: groupings on F by C_T

A graph like that in Figure 6-4 was drawn for the aortic data to construct regions for pooling of cases. Because the numbers of aortas are few, Region 1 was combined with Region 2 and Region 5 was combined with Region 6. NoA aortas showed the expected passage of regression lines near the origin in three of the four data pools (Table 7-3, near zero I_O). The exception was the pool of 10 cases with low C_T and high F. This exceptional pool behaved in the YesA manner, as if some instances of false negative diagnosis may be included in this pool.

Pools of YesA cases have variable outcomes. The two groups with high F, holding altogether 41 cases, show the expected steep slopes and large negative intercepts. The two groups with low F, holding a total of only seven members,

Table 7-3. Means of orthogonal slopes and intercepts relating F to C_T in 172 aortas according to levels of F by C_T; regions as in Figure 6-4.†

Region	Group C_T	Group F	Slope = S_O; µm per cell NoA	Slope YesA	Intercept = I_O µm NoA	Intercept YesA	Number of Cases NoA	Number of Cases YesA
1&2	low	high	32.2	39.0	-236	-206	10	18
3	low	low	5.5	10.0	24	-17	63	5
4	high	high	8.7	** 26.6	48**	-372	24	23
5&6	high	low	4.2	3.8	55	84	27	2

ANOVA	Slope F	Slope (p)	Intercept F	Intercept (p)
Region	26.9	0.006	6.0	0.00
YesA	5.8	0.02	3.3	0.07
Interact	2.7	0.04	7.6	0.00

** = YesA differs from NoA, $p < 0.01$.
†By linear regression, $I_O = -10.4S_O + 91.5$ ($R^2 = 0.75$, n=8); hence $C_T = 10.4$ and F = 91.5 at the region of convergence of the 8 pools of orthogonal regression lines having the average coefficients tabulated here; this falls in the area marked by arrows in Figure 6-2.

resemble NoA cases. (These cases bring to mind the discussion of "false negatives" in Figure 4-1 in Chapter Four.)

The eight versions of Equation 7-2 listed in Table 7-3 give estimates of C_T = 10.4 SMC's per site and F = 91.5 µm at the point of closest convergence (cf. footnote of the table).

7.6. Third way: groupings on C_T by P_A

Figure 7-2 defines six regions for the pooling of cases. The two NoA pools with P_A = 0 (Regions 1 and 2), have shallow slopes and near zero intercepts, as usual (Table 7-4). YesA cases have the expected outcomes in all four pools, whether they display P_A and C_T above or below their respective medians.

Comparison of Figure 7-2 with Figure 6-5 demonstrates the one dramatic failure encountered here for the aortic data to reproduce a detail in the coronary data:

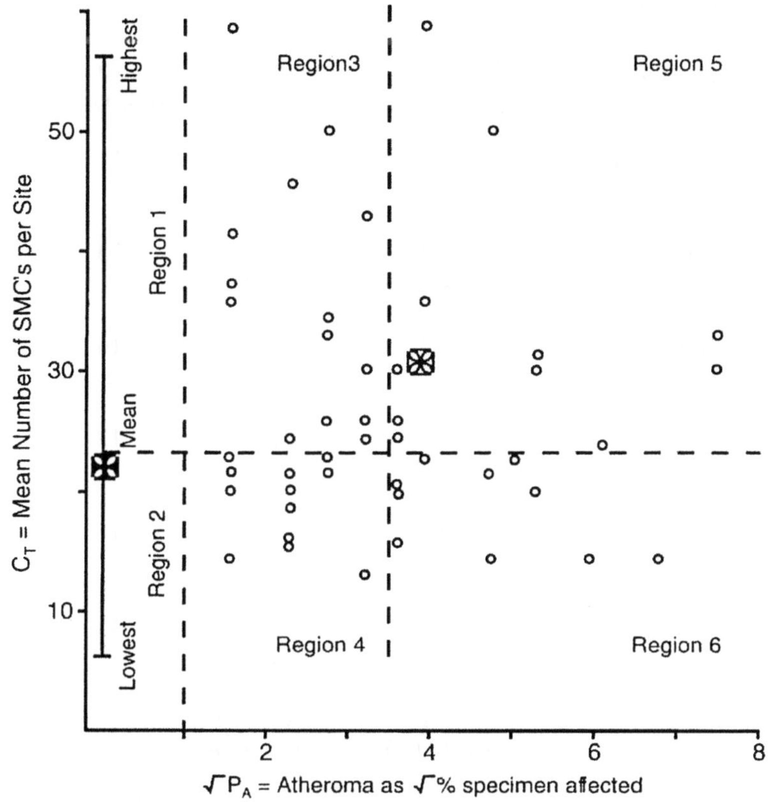

Figure 7-2. Aortic data presented as for coronaries in Figure 6-5, but without significant correlation or regression line within the YesA pool of 48 cases. Numbered regions are used for reference in Table 7-4. Horizontal dashed line is drawn at the median of all cases. Rightmost vertical line marks the median of the YesA cases. Maltese crosses locate means of NoA and YesA pools of cases.

C_T fails to correlate significantly with P_A within the YesA pool of cases. Visual inspection of the graph identifies Region 5 as the source of this disagreement. The upper right corner of the graph contains many coronary artery points, but is empty in the plot of aortic data points. It is as if high values of C_T in one part of the aorta impart resistance to progression of atheroma in other parts of the aorta, whereas such action at a distance fails in the coronary artery. The heterogeneity of arterial architecture in the coronary might explain this discrepancy, as considered later.

Table 7-4. Means of orthogonal slopes and intercepts relating intimal fibroplasia (F) to total SMC's (C_T) in cases grouped by mean C_T and extent of atheroma (P_A) according to regions in Figure 7-2.†

Region	Group C_T	Group P_A	Slope = S_O; µm per cell	Intercept = I_O µm	Number of Cases
1	high	zero	6.3 A	52 A	51
2	low	zero	9.2 A	-12 AB	73
3	high	low	20.9 B	-309 CD	14
4	low	low	31.3 B.C	-179 CD	12
5	high	high	29.7 B.C	-370 D	11
6	low	high	34.1 C	-149 BC	11
ANOVA*			F (p) 16.1 (0.00)	F (p) 10.9 (0.00)	

*Means within a column failing to share a symbol A, B, or C differ significantly at p < 0.05.

† By linear regression, $I_O = -9.5S_O + 47.1$ ($R^2 = 0.48$, n=6); hence $C_T = 9.5$ and F = 47.1 at the region of convergence of the 6 pools of orthogonal regression lines having the average coefficients tabulated here; this falls in the area marked out by arrows in Figure 6-2.

The six versions of Equation 7-2 listed in Table 7-4 give estimates of $C_T = 9.5$ SMC's per site and F = 47.1 µm at the point of closest convergence (cf. footnote of Table 7-4).

7.7. Fourth way: convergences of regression lines for all cases

Table 7-5. Estimated locations for plotting the position on the graph of Figure 6-2 where orthogonal regression lines converge.†

Way of estimation	Location for plotting C_T	F	R^2	N
First				
Coronary	14.3	106	0.82	8
Aorta	13.1	111	0.84	10
Second				
Coronary	10.4	69	0.89	10
Aorta	10.4	92	0.75	8
Third				
Coronary	11.5	77	0.92	6
Aorta	9.5	47	0.48	6
Fourth				
Coronary	9.4	64		
Aorta	13.1	119		
Mean	11.5	86		

† The entries of this table are reproduced from the footnotes of Tables 6-1,2,&3 and Tables 7-2,3,&4, and from the textual passages.

The averages of S_O and I_O were found for each of the two pools, YesA and NoA. For S_O, the averages were 28.5 and 8.0 μm/SMC in the two pools respectively. For I_O, the averages were -254 and 15 μm respectively. Solving these as a pair of simultaneous equations yields $C_T = 13.1$ SMC per site and $F = 119$ μm at the point where the two averaged regression lines cross.

7.8. Comparing the four ways in aorta with coronary

The footnotes of Tables 6-1,2,&3 and 7-2,3,&4 are summarized in Table 7-5, along with the comparable results from the fourth way of estimating convergence zones. The estimates obtained in four ways all cluster around the average of $C_T = 11.5$ SMC/site and $F = 86$ μm. Equally remarkable is the close agreement of values from the aortic data with those from the coronary arteries.

7.9. Estimating cell numbers at censored sites

Chapter Six developed the following formulation, which we can now apply to the aortic data set:

Let C represent cell numbers in rectangular areas such as those in Figure 6-1. C_M denotes SMC's that used to occupy the obscured sites that are *missing* from the observed sample due to censorship by atheromas. C_T denotes the SMC's in the observed part of the specimen, and C_B those in the whole artery *before* censorship. Also let p be the proportion of specimen obscured by atheroma, i.e. $p = P_A/100$. Hence, $C_B = pC_M + (1 - p)C_T$. In a series of n arteries, C_T and p can be measured in each of them, and from the n data pairs, C_B and C_M can be calculated as a and b coefficients in a regression equation, $Y = bX + a$, where $Y = (1 - p)C_T$, $X = p$, $C_M = -b$, and $C_B = a$.

A modification of the Chapter Six approach is needed here, because the correlation of C_T with P_A is not significantly different from zero. Hence the analysis is impractical just within the YesA pool itself. Consequently, the 124 NoA specimens are pooled with the 48 YesA specimens to calculate Equation 7-3 by linear regression

$$Y = -14.9X + 22.8 \; (R^2 = 0.086) \qquad (7\text{-}3)$$

This gives point estimates of 14.9 SMC's/site ($C_M = -b$) as the *missing* cellularity at locations that were censored by atheromas, and 22.8 SMC's as the average cell total ($a = C_B$) presumed to pervade throughout the artery *before* the intrusion of necrosis. These compare with the corresponding values for coronary arteries from Chapter Six, C_M = 15.4 SMC's/site and C_B = 25.8 SMC's/site.

When C_B is calculated for each of the aortas, it is found to average 25.6 SMC's/site in the 48 YesA cases compared with C_T = 27.4 SMC's in uncensored sites at mean age = 55.4 years. The average for NoA specimens of ages 50-59 years is C_T = 25.4 SMC's/site. The biased censoring of low SMC sites by atheroma is fully able to account for the modest elevation of C_T in the observed sites of YesA cases. This result now recurs in the aortas just as we formerly saw in the coronary arteries.

7.10. Estimation of fibroplasia at censored sites

Chapter Six went on to develop the following formulation, which we can now continue applying to the aortic data set. For this purpose, the aortic value of 14.9 SMC's/site is substituted for the coronary value of 15.4 SMC's/site used in Chapter Six:

Equation 6-2 defines the variable F_O, the fibroplastic intimal thickness obtained when an observed regression line intersects the vertical line erected on the abscissa 14.9 SMC/site. The variable $F_S = 14.9F/C_T$ (cf. Equation 6-3) defines the ordinate that obtains when a line emanating from the origin and passing through the means of F and C_T crosses the vertical at abscissa 14.9. In the pool of 124 NoA aortas, F_O = 134 μm and F_S = 126 μm. In the pool of 48 YesA aortas, F_O = 171 μm and F_S = 274 μm.

We can now substitute these values into Equation 6-4 to obtain F_M, our upper limit estimate of the intimal thickness that prevailed at the censored sites before intrusion of atheroma. (This estimation uses the same approximation as in Chapter Six, q = p.) The result is $F_M = 1505$ µm. Again as before, taking $F_S = 274$ µm as the lower limit estimate, we find the range 274-1505 µm for the missing value of F prior to censoring by intrusion of atheroma. This range of values is in reasonably good accord with data in Chapter Three. Table 3-1 indicates average intimal thicknesses for most YesA aortas to fall in the range 200 to 700 µm. Moreover, the estimated thickness of censored sites, F_2, was 45% thicker than the average of observed sites. Hence, the missing values could readily fall in the range 290 to 1015 µm. The further adjustment for atheromatous erosion gives F_3 in the range 490 to 1215 µm. All of these estimates seem consistent with the newly derived values.

7.11. A final check on consistency of reasoning

If the sites obscured by atheroma were those that previously held the greatest amounts of fibroplasia per SMC, then we should anticipate a biasing downward of the observed ratio F/C_T. The maximum possible magnitude of this downward bias is open to assessment by making use of the foregoing deductions. This is done through Equation 6-5, using the aortic value 14.9, $(F/C_T)/(q(F_S - F_O)/14.9 + F/C_T)$, where the observed fibroplasia per SMC, F/C_T, is augmented in the denominator by the quantity presumed to be lost to censorship, $q(F_S - F_O)$ in the atheromatous proportion, q, of the specimen. The average magnitude of the downward bias of the observed F/C_T is 0.71 (compared with 0.88 found for the coronaries). This magnitude is judged to be plausible, and unable to discredit the theory.

7.12. An alternative hypothesis

Chapter Six offered this line of reasoning: "Suppose that sites with few SMC's are resistant to fibroplasia, rather than prone to atheroma. Data points might be missing in Figures 6-2&3, in the region above the rings of arrows, not because they were selectively removed but because they never formed. By this hypothesis, the estimated missing values for C_T would be telling us which sites tended to resist fibroplasia, and F_M telling us how much fibroplasia was resisted in this manner."

Certain details in the coronary data were difficult to reconcile with this hypothesis. The aortic data now raise yet another difficulty: Along the length of the coronary artery, distal sites tend to be consistently less cellular than proximal sites. If these less cellular sites are resistant to fibroplasia by virtue of their distal location, then the uneven anatomy of the coronary artery should exaggerate the phenomenon under discussion, or perhaps even create it artifactually. Yet the lateral thoracic aorta manifests the same phenomenon in the absence of uneven anatomy. Moreover, the magnitudes of estimated parameters in the mathematical models are amazingly similar in the two types of artery. The phenomena, therefore, appear to be independent of location within the highly heterogeneous coronary arterial tree. This independence is hard to reconcile with the alternative hypothesis.

7.13. A speculative synthesis

From age 10 to 25 years, the lateral thoracic aorta grows and develops a normal intima. A substantial random component participates in the dispersal of proliferating SMC colonies, so that SMC numbers vary appreciably from place to place. At an early stage of maturation, the surface is apt to look lumpy in resemblance to cobblestone pavement. When the intima acts to smooth this initially rough surface, it elects not to alter cell numbers, but rather to elaborate collagenous matrix to fill the dales between the hills. This is a general property of arteries in response to stresses and injuries of all sorts, and is of particular interest in aging arteries as they undergo presbyectasia (dilated, tortuous, and rigid), which is typically accompanied by generalized intimal fibroplasia. The sites with few SMC's are the earliest to initiate these events by making excessive matrix, and these localized sites of excess matrix have a large head start when aging later sets in. Evidence for this speculation is considered in Chapter Eight in the discussion of Table 8-5.

7.14. The problem of weak correlations between of F and C_T

A methodological problem arose in the aortic data set that was not a serious concern for the coronary arteries in Chapter Six. Some of the elliptic scatters of data points as in Figure 6-2 became nearly circular. When the correlation coefficient of F with C_T approaches zero, the fitting of orthogonal regression lines becomes erratic, and a few cases can spuriously generate influential outlier values. To avoid this complication, the 18 aortas with $r < 0.21$ were discarded from the analyses in this chapter, and are omitted from the pool of cases used in Table 7-6.

Table 7-6. Descriptive characteristics of selected variables, including correlation coefficients, raw (upper triangle) and age adjusted (lower triangle) in 172 specimens of lateral thoracic aorta.†

					Correlation coefficients						
Variable	Symbol	Units	Mean	SD	S_O	I_O	F	F_C	C_T	P_F	P_A
Age	A	Years	42.7	16.2	.46**	-.30**	.57**	.50**	.23**	.19**	.32**
Slope	S_O	Δμm/SMC	13.7	17.0	.--	-.84**	.67**	.75**	-.01	.24**	.44**
Intercept	I_O	μm	-60.4	266.5	-.82**	.--	-.48**	-.35**	-.12	-.18*	-.26**
Fibroplasia Total	F	μm	255.8	179.4	.56**	-.39**	.--	.77**	.46**	.35**	.65**
Per SMC	F_C	μm/SMC	11.4	8.1	.62**	-.17*	.67**	.--	-.09	.24**	.61**
Cell total	C_T	$N/(100\ \mu m)^2$	23.3	10.2	-.13	-.05	.41**	-.25	.--	.15	.09
Foam cells	P_F	%	13.8	15.2	.18*	-.13	.30**	.17*	.11	.--	.32**
Atheroma	P_A	%	4.3	10.9	.35**	-.18*	.59**	.51**	.03	.28**	.--

** significantly different from zero with $p < 0.01$; * = $p < 0.05$.

†This table contains variables relevant to Chapter Eight; the 172 aortas reported here exclude those with $r < 0.21$, where r is the correlation coefficient between F and C_T, for reasons given in Article 7.14 and in Chapter Eight.

CHAPTER EIGHT

SMC NUMBERS AT VARYING DEPTHS IN INTIMA OF THORACIC AORTA

Abstract. In the negative binomial distribution of SMC's over the lateral thoracic aorta, the parameters of the distribution are governed by the mean and the ratio of variance:mean. Estimates of the variance can be obtained several ways, giving alternative estimates of the parameters. One way to estimate of variance derives from the "within sites" or "error" variance, and this way, in particular, offers evidence that foam cell infiltrates (AHA types IIa and III lesions) mark the transition from aggressive fibroplasia to atheroma.

The intima at each of the 40 sites along the lateral walls of the thoracic aorta can be stratified into multiple levels through its thickness. For convenience, these levels are arbitrarily marked off at 0.1 mm intervals of depth in the inquiry reported here. Since the numbers of SMC's at a site are, by protocol, evaluated in each of the bands of 0.1 mm thickness in the radial dimension passing outward through the vessel wall, a *level* (in italics) is defined to be a square of 0.1 mm by 0.1 mm, i.e. $(100\ \mu m)^2$, and these are numbered sequentially 1, 2, . . . n from the inner most, 1, to the outermost, n. It is a great convenience for our analyses that the averages of SMC numbers at *levels* 1, 2, . . . n do not appreciably differ from each other in any consistent way. Data analyzed here are those reported in reference [63].

8.1. SMC numbers by level in sites of varying thickness

The aortas examined here are described in Chapter Five, Article 5.11, and include an additional nine cases that were formerly discarded, but are retained here. In these 200 thoracic aortas, the 40 sites along the lateral walls yielded 8000 sites for evaluation. Of these, 502 sites contained atheromas, and these were omitted, leaving 7497 NoA sites in the data set. The number of SMC's within a *level* at a site is called C_L. Of the n *levels* at a site, the nth level is usually incomplete, and its C_L is calculated as the ratio of SMC number to thickness of the partial *level.* When the number of *levels* at a site exceeded nine, the tenth and higher *levels* were combined into the ninth by finding C_L as the ratio of SMC numbers to the remaining excess thickness. For construction of Figure 8-1, all sites were combined irrespective of age, YesA status, or other considerations (Figure 8-1 is graphically inexact, because it fails to show the *levels* as square.)

The columns in Figure 8-1 are arranged by increasing intimal thickness at 100 μm intervals. The data in each column are the means of C_L at each of the *level*s for the pool of all sites having the specified thickness. The decline of these numbers from left to right reflects the increasing fibroplasia per SMC that is familiar from previous chapters, because C_L is the inverse of F_C. Some examples of the declining C_L with intimal thickness are shown in Figure 8-2.

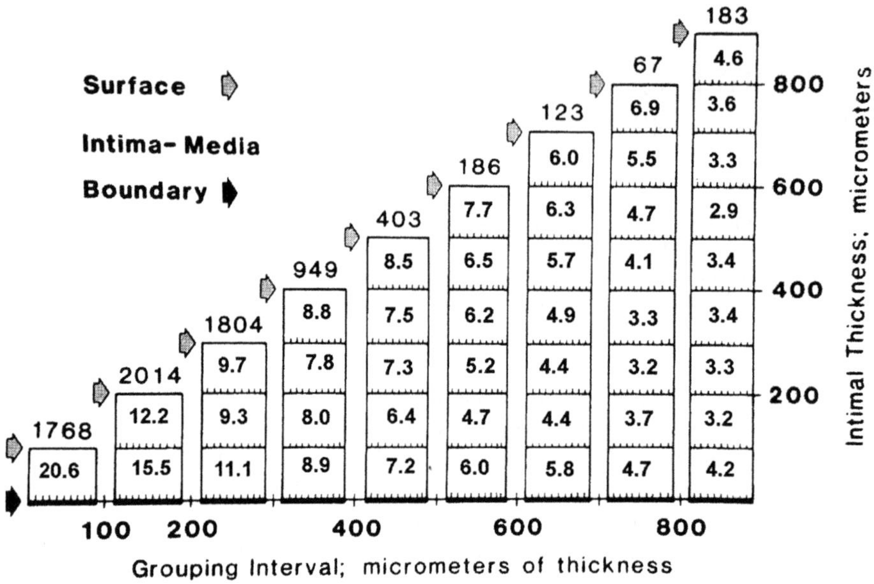

Figure 8-1. Mean number of SMC's within levels of 100 μm thickness at sites of varying total fibroplastic intimal thickness are entered into the rectangles. Numbers capping columns are numbers of sites. Reproduced from reference [63] with permission from Lippincott Williams & Wilkins.

For any given intimal thickness, C_L generally did not differ between *levels*, the exception being a slightly greater C_L in the deepest *level*. The unweighted mean of C_L at the deepest *levels* was 7.93 SMC's compared with 6.15 SMC's at all other *levels* not at the deepest position. This discrepancy might relate to the arbitrary use of a ratio, cells to residual incomplete thickness of partial layers, for calculating C_L. This ratio has a skewed distribution, and its mean may be computationally biased. The discrepancy, although statistically significant, is judged to be negligibly small. Hence, the distribution of C_L over *levels*, at each of the intimal thicknesses, is judged to be sufficiently random to satisfy the needs of sampling theory.

8.2. Influence of age on C_L

The *levels* at each observed site can be safely combined into a pool for averaging, because C_L is acceptably random in its distribution over this sample space; i.e. *levels* at varying depths in a site. Those averages were calculated in 10 year age groups for sites grouped by intimal thickness. The plot in Figure 8-3 depicts age trends for C_L at

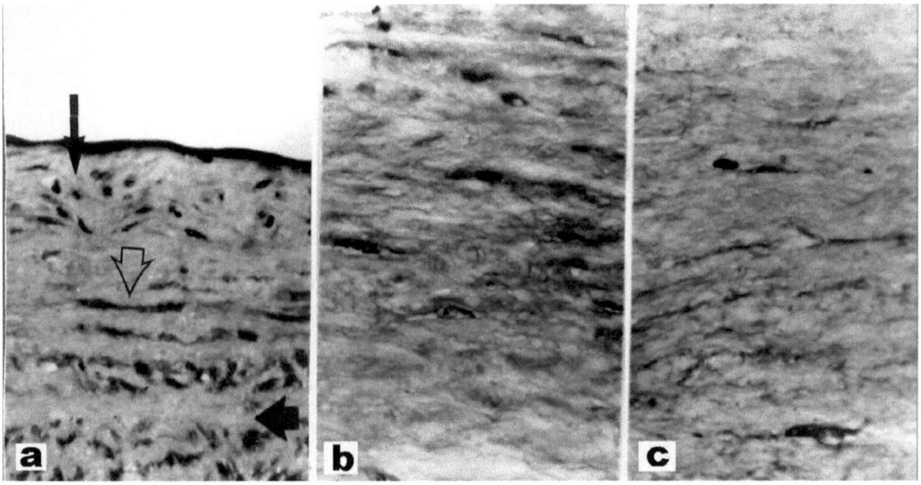

Figure 8-2. Typical cellularities in aortic intimal sites of approximately 100 (a), 300 (b), and 700 (c) μm thickness. The nuclear profiles in the depicted areas number C_L = 28, 6, and 2 SMC's/(100μm)^2 respectively. As in the coronary arteries, most SMC's are longitudinally oriented (white arrow) while scattered media-like islands can be found (black arrow). Black arrow head in "a" marks intima media boundary, and represents 20 μm of length. Modified from reference [63] with permission from Lippincott Williams & Wilkins.

various intimal thicknesses, and these trends reveal a slow, apparently linear decline of C_L as aortas grow older, but only at thicknesses exceeding 200 μm.

8.3. Correlation of C_L with intimal thickness, F, within aortas

Each aorta provides 40 - N sites for evaluation, where N is the number of sites censored by atheroma. Each site provides a mean of C_L averaged over the *levels* at the site. These means of C_L vary inversely with the thicknesses of the sites, thicker sites having lower C_L. The number of SMC's changes little or not at all as a site thickens (the thickening being due to increasing fibroplasia per SMC). Therefore, increasing thickness causes declining SMC numbers within each (100 μm)^2 *level*. This generates a negative correlation between the two variables, C_L and F (Table 8-1).

The 199 aortas of this study were classed into intervals of r, the product moment correlation coefficient of C_L on F. The mean r was -0.34 and did not differ significantly between YesA and NoA cases. The modal value fell in the interval $r > -0.2$ which contains zero. The R^2 assigned to the mean correlation is $(-0.34)^2 = 0.12$. Hence, 88 % of the variance of C_L can be treated as random dispersion over the entire

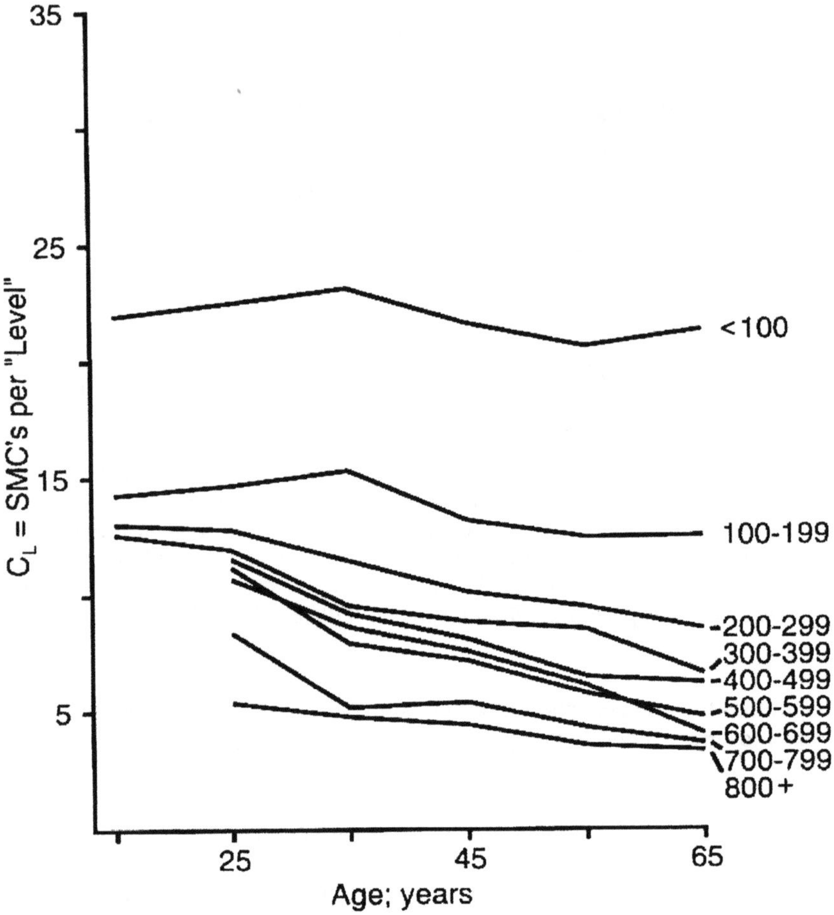

Figure 8-3. Means of C_L are plotted against age. Levels of all depths are pooled within sites grouped by intimal thickness (in units of μm). Angled lines join 10-year age group means.

sample space irrespective of variation of thickness among sites within the aorta. The high value of this quantity suggests that the spreading apart of SMC's by fibroplasia is rather evenly disseminated over the artery wall, which is to say that fibroplasia is generally widespread with only modest effects by focal thickenings.

8.4. Imperfect constancy of realm size, k

Table 8-1. Numbers of cases by correlation of F with the mean of C_L at each site over 40 or fewer sites in lateral thoracic aorta of 199 cases.*

Correlation coefficient	Number of cases NoA	YesA	Total	p-Level†
>-0.2	43	10	53	0.45
-0.2-0.29	27	8	35	0.46
-0.3-0.39	19	10	29	0.48
-0.4-0.49	20	11	31	0.36
-0.5-0.59	18	8	26	0.38
<-0.6	11	14	25	0.34

* Mean r = -0.34; needed for significance is r < -0.30 when n = 40 and p = 0.05 as in most subjects contributing to this table.
† p-level refers to Chi-Square tests of fit to the negative binomial distribution form; this does not associate significantly with the correlation coefficient.

Chapter Five examined the scatter of SMC's over the sites in an artery, by analogy with pennies cast onto a checkerboard. This approach seems reasonable, because the intima is a late addition gained by an artery already formed at an earlier stage of development. Something new is added during adolescent years by developmental migration and proliferation of SMC's, probably originating from the preexisting medial layer. These SMC's could be scattered like pennies, objects dispersed in space. The dispersal of SMC's among the *levels* at a site, however, is a much different process. In this situation, a preexisting population of SMC's newly generate a volume of space for themselves by building collagenous matrix. A variable amount of matrix is newly added to the preformed SMC population. Analysis of this situation would seem to call for treating fibroplasia per SMC as a random variable whose form of distribution is to be sought. Yet, that is not the approach to be followed here, for reasons to be elaborated next:

Let us introduce an assumption that will be revisited later, and develop a line of reasoning from that assumption: Suppose that the SMC's in the arterial intima are not totally immobile within a frozen mass of matrix. Let us propose that SMC's can shift position by as much as 50 to 100 μm over a period of years or decades. (Greater mobility than the specified amount is incompatible with the clonality of SMC's to be considered in Chapter Twenty Five.) This proposal is fully consistent with known facts, and seems entirely credible when viewing tissue sections in the microscope. Only in the fibrous cap and base of necrotic sites do the SMC's seem rigorously locked in place (as "cocoon cells" to be introduced in Chapter Twenty Five).

The mobility assumption just introduced implies that the "realm", as discussed

Table 8-2. Numbers of Chi Square tests comparing negative binomial forms of distribution to observed data on intimal SMC numbers dispersed among *levels* by probability ranges in 186 aortas.*

Probability of Chi Square	Numbers of tests			
	YesA	NoA	Total	Expected
>0.9	8	9	17	18.6
0.8-0.9	4	14	18	18.6
0.7-0.8	4	10	14	18.6
0.6-0.7	6	9	15	18.6
0.5-0.6	1	17	18	18.6
0.4-0.5	6	13	19	18.6
0.3-0.4	5	13	18	18.6
0.2-0.3	4	15	19	18.6
0.1-0.2	7	10	17	18.6
<0.1	16	15	31	18.6

* Chi Square comparing the 10 pairs of total and expected values is 10.45; $0.6 > p > 0.5$.

in earlier chapters, is not a fixed quantity uniquely formed for itself by each SMC. Rather, the community of SMC's are seen to build a neighborhood that they share with each other as they shuffle slowly over short distances. The "realms" are seen as temporary, shifting structures, whose borders blend together, and whose average size can be meaningfully applied to all of the SMC's interchangeably, as done in Chapter Five and later.

The dispersal of SMC's among *levels* at a site is now, according to our mobility assumption, dominated by the movement of SMC's within a preformed space. The problem again becomes one of scattering objects in space. The pursuit of this problem follows next.

8.5. Dispersal of C_L over "levels" and sites

Imagine a specimen composed of 40 sites, with variable numbers of "*levels*" per site. Suppose that SMC's are distributed at random over the whole population of *levels*, irrespective of what site each one occupies. This is, of course, not strictly correct, and must be held as an approximation. We have just seen one approach to estimating the magnitude of the approximation, and that approach obtained 88 % of the variance to be random. For now, let us make the assumption that this approximation is serviceable, and let us follow where it may lead us.

An immediate consideration, of course, is the fitting of negative binomial distribution curves to the scatter of C_L over the *levels*. This seems a reasonable starting place, since that form of curve fits well to dispersal of SMC's over other kinds of sample space. The fitting was done here as in Chapter Five, and again in Chapter Twelve.

The observed values of C_L at all of the *levels* in an aorta were compared with the values expected from the fitted negative binomial distribution by Chi-Square test. The Chi-Squares in turn yielded probabilities, so that cases could be classified by p-values for tabulation (Table 8-2, 5 aortas were unsuitable because of having too many YesA sites, and 8 aortas had only 1 level at all 40 sites, leaving only 186 cases for the analysis). The numbers of cases displaying each of the p-values closely resemble the frequencies expected by chance alone.

8.6. Effects of randomness approximation on curve fitting

In Table 8-2, 16 YesA aortas are found with $p < 0.1$, when only 6.5 are expected. This is a statistically significant excess ($p < 0.01$). Apparently about 1/6 of the YesA aortas fail to fit the negative binomial form for the distribution of C_L. With these few exceptions, readily attributable to biased censoring by atheroma, the remainder of aortas are well described by the negative binomial. This outcome supports our use of the randomness approximation introduced above. The 12 % of variance attributed to the correlation of C_L with F did not cancel the fitting of theoretical curves to the data. This puzzling result calls for some close inspection.

Table 8-1 explores the effect on negative binomial curve fitting arising from the correlation of C_L with F. The table shows p-values for the Chi-Square tests averaged within intervals of r, where r is the correlation between C_L and F. Whether the correlation is weak or strong has no perceptible influence on the fitting of the curves. As sites thicken by fibroplasia, thus elevating F, the resulting dilution of SMC's diminishes C_L, thereby generating the negative correlations of Table 8-1. This effect of fibroplasia has no perceptible influence on the goodness of fit by the negative binomial curves. Hence the approximation introduced above by treating the effect of the correlation as negligible seems workable.

When estimating the parameters of the fitted curves, however, it is important to remember that the approximation could have an impact on the magnitudes of the estimates. The nonrandom effect may be contributing to the shapes of the curves without distorting their underlying form.

(Technical note: The sums of negative binomially distributed quantities are also negative binomially distributed. Hence the nonrandom effect under discussion may follow this form, thereby contributing part of the over dispersion reflected in the parameter c.)

Table 8-3. Means of hypothetical parameters for the logarithmic, M_2, compounded into the Poisson, M_1, to form the negative binomial fit to the distribution of SMC's over *levels* in the thoracic aorta by mean intimal thickness in 61 YesA aortas and in 133 NoA aortas, obtained from V_{LT} (total) or V_{LE} (error) variances.

F; µm	M_1; mean N clusters				M_2; mean cluster size				c-parameter				Number of cases	
Intimal	NoA		YesA		NoA		YesA		NoA		YesA			
thickness	Total	Error	Total	Error	Total	Error	Total	Error	Total	Error	Total	Error	NoA	YesA
<50	8.0	-.-	-.-	-.-	2.6	-.-	-.-	-.-	5.6	-.-	-.-	-.-	8	0
50-99	8.5	11.3	-.-	-.-	2.3	2.0	-.-	-.-	4.4	3.7	-.-	-.-	19	0
100-149	7.9	8.3	6.1	8.8	2.0	1.9	2.2	1.5	3.6	3.3	4.2	2.3	31	2
150-199	6.8	7.1	5.7	6.0	2.0	1.9	2.2	2.1	3.0	3.4	4.0	3.7	13	1
200-249	6.8	7.6	5.3	6.2	1.8	1.6	2.0	1.7	2.6	2.5	3.6	2.6	20	5
250-299	5.5	6.1	5.4	5.9	1.8	1.5	2.0	1.8	2.6	2.7	3.7	3.1	20	6
300-349	4.5	5.0	5.0	6.0	1.7	1.4	1.8	1.5	2.5	2.2	3.1	2.1	15	7
350-399	6.4	7.4	3.9	4.5	1.6	1.4	1.9	1.6	4.2	2.0	3.3	2.6	2	15
400+	5.3	6.0	2.9	3.3	1.6	1.5	1.9	1.7	4.0	2.0	3.4	2.7	5	25

ANOVA	Total						Error					
	M_1		M_2		c-parameter		M_1		M_2		c-parameter	
	F	p	F	p	F	p	F	p	F	p	F	p
YesA	6.1	0.01	3.3	0.09	2.6	0.08	1.9	0.17	0.2	0.71	0.1	0.71
F	4.3	0.00	2.2	0.04	2.0	0.05	6.3	0.00	1.1	0.38	1.2	0.33
Interact.	1.1	0.34	0.0	0.99	0.0	0.99	1.0	0.44	0.3	0.93	0.3	0.95

8.7. Estimating the parameters of the negative binomial fits

Chapter Five introduced the quantities M_1 and M_2 to serve as parameters for fitting the negative binomial distribution curves to each of the data sets. These quantities originate directly from biological considerations. The SMC's represented by C_L are thought to occur in clusters of one, two, three, . . ., in accordance with a logarithmic distribution with mean M_2. The clusters are in turn taken to be scattered in accordance with the Poisson distribution with mean M_1. The distribution of SMC's therefore follows the negative binomial. Equations 5-1&2 tell us that $M_1 = M_L (\log c)/(c - 1)$ and $M_2 = M_L / M_1$, where M_L is the mean of C_L, V_L is its variance, and $c = V_L / M_L$.

Analysis of variance offers us three ways to estimate the variance, V_L, (1) within sites, (2) between sites, and (3) pooled total irrespective of sites. Two of these ways are of special interest, the total variance of the entire data set V_{LT}, and the "error" variance pertaining to the variations among *levels* within sites, V_{LE}. The c-parameters obtained from these two variances have the same denominator, M_L, and are direct reflections of the variances themselves.

Table 8-3 presents the c-parameters for pools of cases within strata of increasing mean fibroplastic intimal thickness. The parameter derived from the error variance fails to differ significantly among thickness groupings or between YesA and NoA cases. Therefore, the term (log c)/(c-1) can be treated as constant. This makes M_1 the reciprocal of M_2, with their variations between subjects determined entirely by M_L. The same cannot be said for the c-parameters obtained from the total variance, because they vary over the thickness groupings ($p < 0.05$) and marginally so between YesA and NoA classes ($p < 0.08$).

8.8. Effects of randomness approximation on parameters

The c-parameters extracted through V_{LT} show differences between variously grouped subjects, while the c-parameters extracted through V_{LE} do not. Hence, c differs between thickness groups and between YesA-NoA classes of cases entirely because of variations <u>between</u> sites in each aorta and not to variation among *levels* <u>within</u> sites. Furthermore, the variation between sites can be entirely attributed to the correlation of C_L with F that was considered earlier. This is shown by examining the ratios of the two c-parameter estimates, error divided by total, V_{LE}/V_{LT}. In14 cases, all sites had only one *level*, thus disallowing estimation of error variance. Excluding these 14 cases the remaining set of 186 suitable aortas held ratios of error to total variance that did not differ significantly among groupings or between classes (data not shown). Their overall average is 0.86, hence, $R^2 = 0.14$. This compares with $R^2 = 0.12$ found earlier by correlation analysis. The two ways of estimating R^2 reach similar conclusions, and therefore seem to be embodying the same biological determinants.

Table 8-4. Means of hypothetical parameters for the logarithmic, M_2, compounded into the Poisson, M_1, to form the negative binomial fit to the distribution of SMC's over *levels* in the thoracic aorta by age in 60 YesA aortas and in 125 NoA aortas, obtained from V_{LT} (total) or V_{LE} (error) variances.

Age;	NoA		YesA		Number of cases	
years	Total	Error	Total	Error	NoA	YesA
	M_1; mean N clusters					
15-29	8.3	9.0	7.0	7.6	35	4
30-39	7.4	9.0	4.0	5.0	28	4
40-49	5.7	6.0	3.9	4.7	21	6
50-59	6.5	7.4	4.0	4.8	24	16
60-69	5.3	6.1	3.1	3.8	17	30
	M_2; mean cluster size					
15-29	1.9	1.9	1.9	1.7		
30-39	2.0	1.7	2.2	1.7		
40-49	1.9	1.8	1.9	1.5		
50-59	1.7	1.5	1.9	1.6		
60-69	2.0	1.8	2.1	1.7		
	c-parameter					
15-29	3.3	3.4	3.1	2.9		
30-39	3.5	2.7	4.0	2.8		
40-49	3.2	3.0	3.3	2.2		
50-59	2.8	2.3	3.2	2.5		
60-69	3.6	3.0	3.8	2.8		

ANOVA	M_1		M_2		c-parameter	
Total	F	p	F	p	F	p
YesA	27.4	0.00	0.9	0.36	0.6	0.46
Age	7.1	0.00	1.4	0.25	1.4	0.23
Interactions	0.5	0.73	0.2	0.95	0.2	0.96
Error						
YesA	14.7	0.00	0.4	0.53	0.6	0.46
Age	3.9	0.00	0.7	0.58	0.8	0.55
Interactions	0.5	0.74	0.3	0.87	0.3	0.88

The differences between the two estimates of c-parameters enter directly into the calculation of M_1 and M_2. Overall, M_1 obtained from the error-derived c-parameter is 1.16 times that of the total-derived c-parameter. M_2 is 0.90 times as great. These proportions do not differ between groupings of intimal thickness (data not shown). They do, however, differ significantly between classes: the ratio for M_1 averages 1.10 in NoA and 1.21 in YesA classes respectively while the ratio for M_2 averages 0.92 and

0.84 respectively.

(Technical note: The significant YesA effect just considered is the expected result from biased censoring of low SMC sites by atheroma. The censoring caused broadening and flattening of the distribution curve for SMC's in Chapter Five, Figure 5-1, which is to say raising the ratio, $c = V_L / C_L$.)

8.9. Constancy of the c-parameter across age groups

Table 8-4 indicates the constancy of c-parameters across age groups and between YesA and NoA classes. This was so whether the c-parameters were extracted from V_{LT} or V_{LE}. The constancy of the c-parameter forces a similar constancy onto M_2, which is also indicated in Table 8-4.

8.10. Importance of variations between sites

Table 8-3 contrasts sharply with Table 5-4 in Chapter Five. The c-parameters showed variation across thickness groupings and between YesA-NoA classes in Chapter Five. Such variation as found in this chapter is confined to V_{LT} derived parameters and is absent in the V_{LE} parameters. Even the significant variations found here in V_{LT} derived parameters are greatly reduced from those in Chapter Five.

The variable under study in Chapter Five was C_T, total SMC numbers at each site summed and not averaged over *levels* as for C_L. The variation was entirely between sites, without consideration of *levels* within sites. It seems plausible to suggest that the main source of that variation is uneven formation of normal intima during organogenesis. The seeding of SMC's into the newly forming intima of childhood is not entirely homogenous. Some sites receive many SMC's while others receive few. In the lateral thoracic aorta the uneven seeding of SMC's is random, because no site is especially prone compared to all others.

We have reasons to think that the sites with few SMC's begin early to form excessive fibroplasia per SMC. The reasons will be reviewed shortly. The evidence suggests that the SMC's in the sites with few cells try to smooth the uneven intimal surface by building up the fibroplastic matrix. This activity dilutes the cells with increased matrix per cell. If this is true, then the SMC's or their excessive matrix at these sites begins the aging process much sooner than at other sites. When age later asserts its effect by aggressively expanding the fibroplasia toward the 9.3 μm threshold for atheroma, the sites with few SMC's seem to have a long head start in this process.

8.11. Plots of F by C_T often show an upward bias in the intercept terms of regression lines

Table 8-5. Foam cell infiltrates, numbers of SMC's, orthogonal intercept parameter, and c-parameter according to fibroplastic thickness of intima.*

F; μm Intimal thickness	Foam Cells P_{Fi} %	SMC's; N C_T	Intercept; μm I_O	c-param	Number of cases†
		--- NoA Cases ---			
<50	5.8AB	8A	22A	2.1A	7
50-99	9.0AB	15A	28A	3.0A	23
100-149	6.5A	19B	39A	3.4AB	28
150-199	13.1AB	25C	69A	2.8A	13
200-249	9.6AB	26C	8A	3.3AB	17
250+	13.9B	28C	-31A	4.3B	36
		--- YesA Cases ---			
All	23.2C	27C	-253B	6.1C	48

* Means within a column that fail to share a symbol A, B, or C differ significantly, $p < 0.05$.
† The 172 aortas of this analysis are described in Table 7-6.

The reasoning begins with graphs like those in Figure 6-2, and continues the development begun in Chapter Seven. The scatter plot of F by C_T is fit by an orthogonal regression line for each aorta. The coefficients of the equation are the slope and the intercept, S_O and I_O respectively. These two variables for the aorta generally tend to recapitulate the pattern for the coronary artery. Slopes are shallow for the NoA aortas but steep for the YesA specimens. Intercepts are generally near the graphic origin for NoA aortas but strongly negative for the YesA specimens.

Unlike the coronary arteries, however, the NoA aortas reveal a consistent trend toward upward bias in the I_O values. Nearly 70 % of the I_O values are positive, and their average is positive in five of the six groups of NoA cases in Table 8-5. The lines tend to pass near the origin, but the majority are clearly above the origin. Hence, F is not, as a rule, directly proportional to C_T. Instead the ratio usually rises as C_T declines.

The upward bias in I_O tells us that fibroplasia per SMC tends to rise as SMC numbers fall. The pattern is already established in the group of specimens with the thinnest intima, where $I_O = 22$ μm when $F < 50$ μm. Of the 35 NoA subjects of ages 15 to 29 years, 33 have $I_O > 0$ (data not shown). The median of I_O in these youngest subjects is 37 μm, and its mean is 48 μm which is significantly different from zero ($p<0.0001$). This is the evidence that we were seeking: Sites with the fewest SMC's appear to be the earliest sites to initiate fibroplasia per SMC and to continue thereafter leading the advance of further intimal thickening over time. This trend would, of course, bring these sites to the threshold for atheroma sooner than other sites.

The pattern seems to fail, however, when thickness of intima surpasses 200 μm. The NoA specimens then show a trend toward a resemblance to the YesA condition. Average SMC numbers increase, I_O declines into the negative region, and, most intriguing, the c-parameter increases. These YesA traits were earlier credited to biased censorship of low SMC sites by atheroma. Clearly that explanation cannot be invoked when atheromas are absent. The paradox calls for an explanation.

8.12. Biased censoring of low SMC sites by fatty streak elements

The variable P_F refers to the percentage of nonnecrotic specimen occupied by fatty streak elements, as evidenced through the finding of foam cell infiltrates in the tissue sections. These foam cell collections obscure the assessment of SMC numbers, just as atheroma does. The effect is often on a small scale, so that a substitute position can usually be found within 100 μm of the formally designated location. These substitute positions are used extensively in practice. It therefore seems possible that the substitute positions could be biased with respect to SMC numbers.

Consider a diagram like Figure 7-2, constructed with C_T plotted against P_F. This introduces the measure of foam cell infiltrates, P_F, in place of the measure of atheroma, P_A, shown in Figure 7-2. A version of Equation 7-3 can be constructed appropriate to the envisioned diagram: Set $Y = C_T(1 - p)$, and $X = p$, where $p = 0.01P_F$. Fit a regression equation of the form $Y = bX + a$. Doing this yields values of b that do not differ significantly from zero in the pool of YesA cases or in the pool of NoA cases. Statistically significant results in the pool of 172 YesA and NoA cases combined were $b = -16.5$ and $a = 22.2$ ($R^2 = 0.077$).

The numbers of SMC's at locations censored by foam cells is estimated to be 16.5 cells/site. The average of all obscured plus observed locations is estimated to be 22.2 cells/site prior to the censorship. (A similar analysis of the coronary arteries, using data from Chapter Five, obtains $b = -19.3$ and $a = 25.1$ ($R^2 = 0.097$).) This outcome implies that foam cells have a small propensity to select locations with fewer SMC's than average. This could mean that the transition from fibroplasia to atheroma passes through a stage of infiltration by fatty streak elements, since the same sites are at risk of both fatty streak and atheroma features.

8.13. The transitional lesion; from fibroplasia to atheroma

In Chapter One, Figure 1-1D illustrates the coexistence of fatty streak elements with aggressive fibroplasia ($F_C > 9.3$ μm). The American Heart Association Lesions Committee refers to this appearance as type IIa or type III lesion. The type III lesion is often proposed to be a transition between the lipid infiltration-induced type IIb (fatty streak) and type IV (atheroma), where type IIb is illustrated in Chapter One by Figure 1-1C. That kind of transition is what happens in laboratory animals such as the cholesterol fed rabbit. We saw evidence for its occurrence in some human coronary arteries in Chapter Four (cf. discussion of Figure 4-1). However, we have reason now to suggest that this transition does not pass by way of the type III lesion. Rather, it likely passes directly from thin fatty streak (Figure 1-1C) to thin atheroma with subsequent proliferation of a fibrous cap taking it to the type V stage (not illustrated in Figure 1-1).

The transition envisioned here is the much more commonplace change from age induced aggressive fibroplasia to atheroma by way of lipid sequestration. Table 8-5 now gives us further evidence for this proposal. The 36 NoA specimens with much fibroplasia, those with $F > 250\ \mu m$, display resemblances to YesA specimens in I_O, C_T, and c-parameter. Those same specimens also display elevated P_F. It is therefore possible to propose that biased censorship of low SMC sites by foam cell infiltrates can account for the resemblance to the YesA condition.

The result strengthens the statements in Chapter Four (cf. Equation 4-7 & Figure 4-3). The aortic data of this chapter serve to reproduce the coronary results given in Equation 4-7.

$$W = 1.15F_C + 0.70P_F\ (T^2 = 1.61\ \% = 74\ c = -2.63) \qquad (8\text{-}1)$$

W is the weighted average that optimally distinguishes YesA from NoA aortas when $W = 0$. This discriminant function is drawn from subjects of ages 35 to 79 years (F_C and P_F have respective standard deviations of 9.3 and 14.3.) (Symbols are detailed in Chapter One, Article 1.7.)

In the coronary data, Equation 4-7 could "explain" in full the statistical effects of age upon atheroma. That is not true here. Age, when offered, contributes significant additional information in Equation 8-1. The matter calls for further investigation to clarify the discrepancy. However, the placing of fatty streak elements, P_F, on the route to atheroma seems confirmed here.

8.14. The mobility assumption revisited

Earlier in this chapter, an assumption was introduced to foster the development of theory. It was assumed that slight mobility of SMC's over many years causes the random dispersal of these elements in the intimal space. The justification for that assumption can now be reviewed. Flowing from the assumption is a remarkably consistent and quantitatively precise edifice of theory. The theory has several valuable properties: (1) It has so far encountered no serious contradictions from any aspect of the data. (2) It offers resolutions to several longstanding paradoxes, including the geographic paradox that populations alike in extent of fatty streaks in youth later diverge widely in raised lesions at older ages (cf. Chapter Sixteen). (3) It generates useful predictions about findings to be expected from future inquiries. The assumption of SMC mobility within the confines of a small region of intimal volume therefore seems worthy of retention.

CHAPTER NINE

HISTOLOGIC APPEARANCES OF SMC CLUSTERS AND REALMS

Abstract. Under special circumstances, the clustering of intimal SMC's into layers can be demonstrated by connective tissue stains that reveal the lamellae of condensed reticulin between layers. Such stains disclose progressive dissolution of the layers with aging. The dissolution may be important for the construction of lipid traps at sites where solitary SMC's build excessively bulky collagenous realms for themselves.

Chapters One and Five introduce the concepts of fibroplasia per SMC and of SMC realm, two variously obtained estimates of the same quantity, the volume of space occupied by the average SMC. The data of Figures 1-3 and 5-2, in Chapters One and Five, demonstrate the profound importance of this quantity for the intrusion of atheroma. Indeed, this quantity is proposed here as the diagnostic criterion for the disease entity, fibroplastic arteriosclerosis. Unfortunately, this quantity is invisible or blurred most of the time in tissue sections. It is generally not accessible to direct viewing and remains a hypothetical entity in most circumstances. Nevertheless, some value might be gained by looking at the microscopic slides.

9.1. Initiation of SMC clusters during early growth

The intimal layer of arteries undergoes rapid growth during the years of adolescence. Smooth muscle cells multiply rapidly in numbers and later build increasing volumes

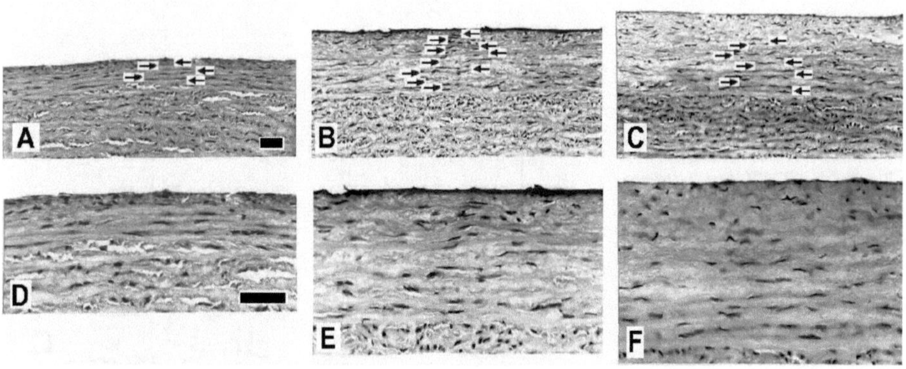

Figure 9-1.Representative findings are shown as seen in an H&E stained paraffin section of coronary artery from a 21 year old man.Arrows mark pale staining zones between darkly staining SMC's in layered clusters. Frames D, E, and F are magnifications of Frames A, B, and C respectively. Bars = 50 μm.

of matrix materials around themselves. At first, the newly forming intima often resembles medial architecture, with the closely spaced SMC's oriented circumferentially. Later the intimal SMC's rotate into a longitudinal orientation along the direction of the arterial axis. This axial orientation often becomes the most easily recognized feature of the intima-media boundary, especially in the aorta, because the internal elastica generally breaks up and displays intermittent duplications. Residual islands of media-like architecture are surprisingly frequent, made of circumferential SMC's, usually clustered in thin layers that can be at any depth from the luminal surface, but most often persist in the deepest intimal layers (sometimes called "musculoelastic layer", a term that fosters confusion).

During growth, the layered architecture of the intima is often boldly conspicuous (Figure 9-1A,D).As the extracellular bulk increases, the layered pattern becomes increasingly obscure, especially in its superficial levels (Figure 9-1 B, C, E, F).In the H&E stains of Figure 9-1, darkly stained layers occupied by SMC's are separated by pale zones at roughly constant intervals through the intimal thickness (arrows).Silver stains for reticulin reverse the color pattern (Figure 9-2); the unstained layers of cells are separated by black intervening zones (arrows).As the aging intima thickens, the condensed reticulin network between layers often persists after the layering of SMC's fades. Eventually, even these remnants of former layers blur and vanish.

9.2. Thicknesses of SMC layers and reticulin lamellae

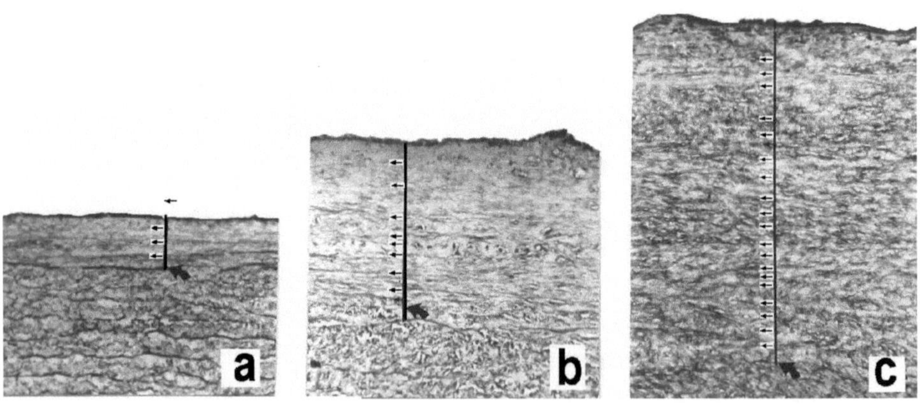

Figure 9-2.Silver reticulin stains of paraffin sections display intima of a youthful aorta, a, and two middle aged aortas, b and c. Vertical lines mark the intimal thicknesses. Arrows mark the darkly stained layers of compacted reticulin that correspond to the pale zones in Figure 9-1. Reproduced from reference [62] with permission from Springer-Verlag.

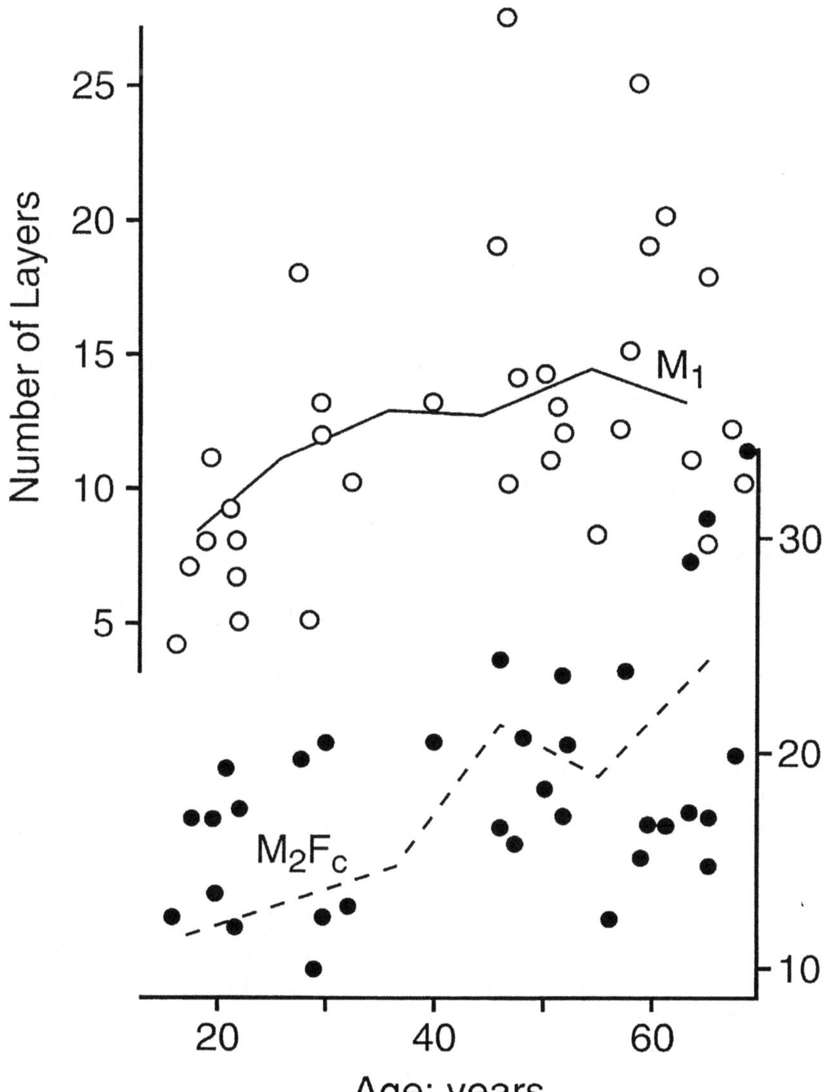

Figure 9-3. Average numbers (circles, scale left) and thicknesses (dots, scale right) of intimal layers, as exemplified by pairs of arrows in Figure 9-2, are shown for 33 typical aortas. Angled solid and dashed lines plot data from Table 5-2 in Chapter Five. These lines are not fitted to these data points, but represent theoretic quantities to be compared with the data. Reproduced from reference [62] with permission from Springer-Verlag.

The arrows in Figure 9-1 have an average distance between them of 12.6 μm. The arrows in Figures 9-2a,b,c respectively have average separations of 15.0, 24.0, and 21.4 μm. These photographs illustrate that lamellae can sometimes be counted in aortic intima. The ratio of intimal thickness to the number of lamellae then determines the average thickness of the inscribed layers. Numbers of layers and their average thicknesses were determined in this way for a collection of representative aortas (cf. Chapter Five, Article 5.7), and the results are plotted in Figure 9-3.

The circles representing the average number of intimal layers rise in youth to a nearly level plateau that is sustained through ages 25-69 years, implying that the construction of new layers ceases with the cessation of growth and maturation. The dots representing thickness of layers pursue an upward course throughout adult life, following a trail that seems vaguely sigmoid. The steepest rise seems to be from ages 30-45 years, with slackening of growth after that age. This is our first encounter with this sigmoid shape, which recurs with ever increasing importance in later chapters.

9.3. What reticulin lamellae tell us about SMC clusters

Overlying the circles in the scatter plots of Figure 9-3 is an angulated line that plots the mean values of M_1 taken from Table 5-2 of Chapter Five. Overlying the dots is a dashed line that plots the product term, $M_2 * F_C$, taking these values also from Table 5-2. M_1 was previously calculated as the mean number of realm *clusters*, and $M_2 * F_C$ as the average thickness of SMC *clusters*. Hence, the layers visible in tissue sections, such as those illustrated in Figures 9-1&2, appear to correspond to numbers and thicknesses of *clusters* rather than individual SMC's. These coincidences lead to an interesting conclusion: $M_2 * F_C$ is a measure of *cluster* thickness in the radial dimension, passing outward through the wall thickness. The *clusters formed of realms joined end-to-end along the arterial axis* form layers that have thicknesses as observed in intimal layers inscribed by the reticulin lamellae.

With these considerations in mind, we can use the data of Table 5-2 to guide our inspection of Figure 9-1. The two lowest arrows in Frame A, for instance, delimit a region that seems to contain nuclei of three SMC's joined end-to-end to form a layer (better seen in the magnified image of Frame D). Hence the 12.6 μm of thickness between the arrows is shared by 3 SMC's giving 4.2 μm as the average "realm". The next pair of arrows delimit two SMC's so that "realm" is determined as 6.3 μm. Results are similar to these for all adjacent pairs of arrows in Figure 9-1. This result is in keeping with the finding in Table 5-2 of $M_2 \approx 2$ SMC's per cluster and with the finding in Table 5-3 of realm ≈ 4 to 6 μm/SMC in the aortas with little intimal thickening, as they are seen to be in the photos shown here. It is informative that the data for aortas in Chapter Five apply so well to the photographs of coronary artery in Figure 9-1.

The clustering of SMC's within layers, typically end-to-end in twos and threes, seems visually to span distances consistent with the value 119 μm. The left-right

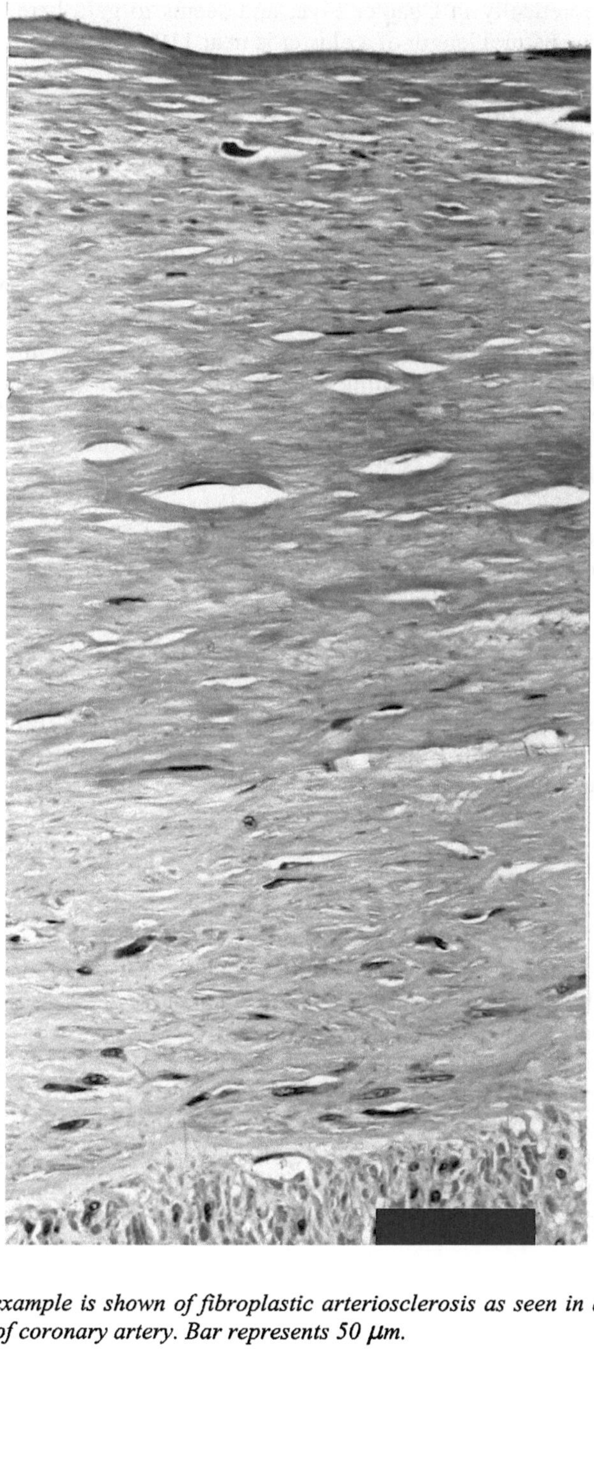

Figure 9-4. An example is shown of fibroplastic arteriosclerosis as seen in an H&E stained paraffin section of coronary artery. Bar represents 50 μm.

distance along the axial dimension of the aorta often appears near this value. This value was obtained theoretically in Chapter Five, and seems to gain here some concrete reality. Perhaps the natural length of a cluster is near 119 μm.

9.4. Fibroplastic arteriosclerosis

Figure 9-4 illustrates the putative entity, fibroplastic arteriosclerosis, because the fibroplasia per SMC, F_C = 25.3 μm/SMC, exceeds the often observed 9.3 μm/SMC estimate for the cutoff value of normal. Most of the intimal area is occupied only by solitary SMC's widely separated from each other. Clusters of 2, 3, or 4 SMC's arranged end-to-end in vaguely defined layers are seen, mostly in the deepest levels of intima. The kind of intima seen in Figure 9-4 evolves from the kinds seen in Figure 9-1. The evolution consists of SMC's elaborating great quantities of collagenous matrix around themselves. Most of this elaboration seems to attend the solitary SMC's with less participation by the clusters. All of these properties are predicted by the theories of Chapter Five. Although this type of intima is prone to sequestering perifibrous lipid, nevertheless, frozen section fat stains often reveal absence of lipid, even though lipid staining is intense in other parts of the same slide, giving clear evidence that the lipid deposition is a consequence and not a cause for fibroplasia of this kind.

9.5. A speculative synthesis

The photographs examined here bring to mind some intriguing results reported by Rekhter et.al. [41]. Those investigators observed interconnections between SMC's in the intima of young, healthy aortas, in the form of filamentous extensions often joining SMC's by gap junctions, mostly within layers and less so between layers. Those connections were broken in the sites infiltrated by fatty streak elements, as if the clusters of SMC's were dispersed into solitary units. Those results might relate to the discussion in Chapter Five.

Evidence in Chapter Five led to the proposal that atheroma tends to favor intimal sites with exceptionally large realms, occupied for the most part by solitary SMC's, while sparing sites with clusters of two or three or more SMC's. The photos in this chapter now illustrate the general finding that aging aortas progressively dissolve the layered pattern of the intima, obliterating the previously visible boundaries between adjacent layers. This unbinding of the layers, accompanied by dissolution of SMC clusters, could be a critical step in the altered behavior of aging SMC's. Those aging SMC's build realms of excessive bulk, probably made of abnormal kinds of matrix materials. Perhaps they do this after losing their interconnections with each other.

CHAPTER TEN

DIRECT IMAGING OF THE HYPOTHETICAL QUANTITY, SEQUESTERED LIPID

Abstract. The quantity, sequestered lipid, entered all previous chapters as an invisible hypothetical entity. In this chapter, the quantity is made visible by use of frozen section fat stains, which reveal lipids to be gauged in intracellular, I_C, and extracellular, E_C, compartments, where E_C is taken to be the measure of sequestered lipid. Predictions from the sequestration hypothesis are upheld with a high degree of precision in all circumstances.

This chapter departs from prior practice by introducing a different way of staining the histologic sections of lateral thoracic aorta. Previously, H&E stains (and connective tissue stains for selected purposes) were applied to paraffin sections. That method offers limited access to visualizing lipid deposits. Only the visible "fatty streak elements", including chiefly macrophage foam cells and their attendant lymphocytes, provide information about lipid trapping. The present study makes use of fat stains upon frozen sections, thereby revealing all "lipids", in the sense of substances that take up the lipid soluble dye, Oil Red O.

The difficulty of cutting frozen sections limits the amount of tissue that can be processed. This in turn precludes the effort to assess the extent of involvement throughout the specimen. Furthermore, the quantities of lipid in visible deposits cannot be accurately measured and must be assessed in rough gradations subjectively evaluated. Despite these limitations, however, such subjectively acquired information does enable relative comparisons from place to place in the specimen, such as at the various depths into the intima and in fibroplastic thickening of variable degrees from one site to another.

In this chapter, data are assembled from the lateral thoracic aortas in a set of 123 cases drawn from an international collection of aortas [65] also used in Chapter Five. Details of the study can be found elsewhere [64]. The formerly hypothetical quantity, sequestered lipid, here measured by the variable E_C, becomes a direct object of attention. The sequestration hypothesis can therefore be confronted with rigorous challenges, which must be passed in order to sustain the viability of the hypothesis.

10.1. Methods

Samples of 2 cm length at the 6th and 10th intercostal levels on the left and right sides of thoracic aorta were submitted to Oil red O staining of 20 μm frozen sections. Stained sections were marked at the middle and two ends to define intervals of 1 cm, 12 marked positions in each aorta. YesA status was taken from Chapter Five evaluations of H&E stained paraffin sections. Under the 10X objective lens, an eyepiece grid marked the intima into layers of 100 μm thickness and 1 mm length; these areas are called "microregions". The deepest layer was treated as whole if over one-half of its area was

Table 10-1. Descriptive statistics for 5 variables measured in 3169 microregions in 123 aortas, combining men and women of all ages and causes of death including those with atheroma (NoA and YesA aortas pooled).†

Variable	Symbol	Units	Mean	SD	Correlation coefficients * θ	D	T_C	E
Age	A	Years	50.2	15.2	0.46	0.32	0.35	0.41
Thickness	θ	N–Levels	3.69	2.41	-.---	0.69	0.43	0.41
Depth	D	N-Levels	2.35	1.73	0.65	-.---	0.37	0.50
Total stain	T_C	Bu	1.91	1.53	0.32	0.29	-.---	0.79
Extracellular	E_C	Bu	1.44	1.32	0.27	0.43	0.75	-.---
Intracellular	I_C	Bu: This is the difference, $I_C = T_C - E_C$, using group means.						

*Upper right quadrant gives raw coefficients; lower left quadrant gives partial coefficients adjusted for age. All coefficients are significantly different from zero, $p < 0.001$.

† Reproduced from reference [64] with permission from Springer-Verlag.

occupied by intima at the intima-media junction; otherwise it was treated as absent. The entire area of each microregion was compared to an array of photographs representing a scale of staining intensity of 0 to 5, and the nearest match was recorded. For want of a name, these units of measure are called "Brilliance Units". After recording the total staining intensity, call this T_C, the proportion of stain judged to represent extracellular lipid was recorded to the nearest 0.1. The product of the total intensity times the extracellular fraction is the extracellular severity grade; call this E_C. All sites with atheroma were excluded from the grading. A total of 3169 microregions were observed at the 12X123 = 1476 evaluated sites (excluding YesA sites).

10.2. Description of variables

This is given in Table 10-1. Brilliance of total stain, T_C, strongly correlates with brilliance of extracellular stain, E_C, $r = 0.790$). This strong correlation is in part computational, because E_C is calculated as a fraction of T_C. It is also partly biological, because lipid laden cells tend to accumulate around extracellular lipid deposits. The

Table 10-2. Mean intensity of ORO staining in 3169 microregions from 123 aortas by age, pooling data from all levels in the intimal depth.†

Age; years	ORO intensity; Bu * T_C	E_C	I_C	Numbers of Regions	Levels
10-19	0.89	0.47	0.42	72	1.0
20-29	1.39	0.75	0.65	357	1.3
30-39	1.31	0.89	0.43	399	2.3
40-49	1.47	0.99	0.48	398	1.9
50-59	1.94	1.57	0.47	787	2.4
60-69	2.56	2.03	0.53	1156	2.9

* T_C = Total, E_C = Extracellular, I_C= Intracellular; I_C is calculated as $T_C - E_C = I_C$
† Reproduced from reference [64] with permission from Springer-Verlag.

two variables contain much information in common, and they can generally substitute for each other in multivariate analyses. Depth of a microregion, D, is D = 1 at distances of 0-99 μm from the luminal surface, D = 2 at 100-199 μm, . . ., and D = 9 at 800+ μm. D correlates strongly with intimal thickness, θ (in units of number of layers as for D), r = 0.696. Again, each of these variables can often displace the other in multivariate analyses.

10.3. Trends on age

Intensity of total ORO staining, T_C, and of extracellular staining, E_C, generally tended to increase with age (Table 10-2). For total staining measured in brilliance units (Bu), the rise was from 0.89 Bu to 2.56 Bu from ages 10-19 to 60-69 years. Estimates for extracellular staining indicated a comparable rise from 0.47 Bu to 2.03 Bu. Taking the difference between the objectively assessed total staining and the roughly approximated estimate of extracellular staining (using group means) as a reflection of lipid-rich cells, the estimated intracellular stain, $T_C - E_C = I_C$, was found to show a generally unchanging trend with age; this is consistent with the insignificant correlation of P_F with age, r = 0.04, seen in Chapter One, Table 1-1a. Intimal thickness, measured in numbers of levels 100 μm thick, tended to increase with age (Table 10-2).

10.4. Trends on intimal depth

The intensity of staining increased with increasing depth into the intima, after exclusion of sites with atheronecrosis (Table 10-3). For total staining, the increase was

Table 10-3. Mean intensity of ORO staining in 3169 microregions from 123 aortas by depth in the intima, pooling data from all ages.*†

Depth; N-levels	ORO intensity; Bu T_C	E_C	I_C	Numbers of Microregions
1	1.47	0.92	0.55	1309
2	1.84	1.32	0.52	825
3	2.23	1.77	0.46	427
4	2.43	2.08	0.35	235
5	2.73	2.40	0.33	142
6	3.13	2.96	0.17	93
7	3.60	3.45	0.15	65
8	3.60	3.44	0.16	41
9	4.13	3.90	0.23	32

* Footnotes as in Table 10-2.
† Reproduced from reference [64] with permission from Springer-Verlag.

from 1.47 Bu at depths of < 100 μm (level 1) from the intimal surface to 4.13 Bu at depths of 800-899 μm (level 9). Estimates of intracellular staining, reflected in the difference between total and extracellular, declined from 0.55 Bu at level 1 to 0.23 Bu at level 9. The deeper levels tended to be from somewhat older aortas than were the superficial levels (data not shown).

10.5. Interaction of thickness and depth effects

The simultaneous actions of age, thickness, depth, and YesA condition cannot be approached by subgrouping of observations, because numbers of observations are insufficient. The matter can be approached, however, using multiple regression. This must be done cautiously, of course, because multicolinearities arise when intercorrelated variables share information; and the variables should not be treated as independent signals about biological processes. The approach taken here is to fit a complete equation, and then to attempt simplifying it by eliminating variables of least importance.

10.6. Deconstructing the regression equations

With $Z = T_C$ and $Z = E_C$, equations were determined with Z as a function of various

Table 10-4. Coefficients of multiple regression are given for selected combinations of regressor variables and their interactions with two choices of Z in the equation Z = f(A, θ, D, Y).

Equation name:	10-1	10-2	10-3	10-4	10-5	10-6	10-7	10-8
Covariates and their interactions	Coefficients for Z = Total lipid				Coefficients for Z = Extracellular lipid			
	Full equation	Excludes θ	Excludes Y	Excludes Y and θ	Full equation	Excludes θ	Excludes Y	Excludes Y and θ
Intercept	2.11	1.19	2.07	0.66	1.26	0.73	1.30	0.32
Age (A)	-0.0251	0.00047	-0.0177	0.0143	-0.0127	0.00016	-0.0098	0.0088
Thickness (θ)	-0.436		-0.034		-0.316		-0.240	
Depth (D)	-0.601	-0.428	-1.330	-0.124	-0.492	-0.426	-0.792	-0.166
YesA (Y)	-0.551	-0.131			-0.196	-0.301		
A θ	0.0134		0.00346		0.00726		0.00434	
θ D	0.093		0.157		0.063		0.114	
AD	0.0119	0.0127	0.0265	0.0068	0.0131	0.0141	0.0202	0.0086
a θ D	-0.00177		-0.00282		-0.00101		-0.00205	
AY	0.0310	0.0144			0.0183	0.0121		
θ Y	0.806				0.094			
YD	-0.674	0.572			0.179	0.572		
YA θ	-0.0191				-0.00510			
YAD	0.0120	-0.112			0.0022	-0.0112		
Yθ D	0.0186				0.0016			
R^2	0.281	0.244	0.251	0.217	0.367	0.355	0.351	0.345

combinations of these four regressor variables: A, θ, and D as in Table 10-1, and Y, where Y = 1 for YesA specimens and Y = 0 for NoA specimens. The interactions of these variables with each other are introduced by taking their products. Table 10-4 gives the two full regression equations, 10-1 and 10-5. Attempts were made to simplify these models for T_C and E_C by eliminating one of the variables A, θ, D, or Y. For Z = T_C, the values of R^2 were 0.232, 0.244, 0.257, and 0.251 when removing A, θ, D, or Y respectively (down from 0.281 in the total equation). For Z = E_C, the comparable outcomes were R^2 = 0.296, 0.355, 0.262, and 0.351 (down from 0.367). For Z = E_C, removal of the terms in A or D, but not θ or Y, caused serious deterioration in R^2, showing that a simplified model would require A and D but not θ or Y. For Z = T_C, each of the same two variables, θ or Y, could be chosen for removal with little loss of R^2.

Reduced equations found by eliminating Y, θ, or Y and θ, are given in Table 10-4. Plots in Figures 10-1,2&3 use values calculated for T_C and E_C through Equations 10-2 and 10-6 which remove θ and retain A, D, and Y; Figures 10-1&2 result by using Y = 0 in the equations, and Figure 10-3 by using Y = 1. Values calculated for E_C and T_C through equations 10-3 and 10-7, which eliminate Y terms, are given in Tables 10-5 and 10-6 respectively.

10.7. ORO stains in relation to age and intimal depth in NoA aortas

From ages 15 to 35 years average intimal thickness increases in these data, but with negligibly few locations thick enough to offer observations beyond the depth of level 4. Total ORO staining is about 1 Bu of intensity at age 15 years, and extracellular staining is about 0.7 Bu (Figures 10-1&2). The intensities range around constant levels for the 20 years of aging from 15 to 35 years, and are generally similar at all depths, as the thicker sites begin to emerge. Only four Yes-A aortas were under age 40 years, which justifies restricting attention to the data for NoA specimens.

After age 40 years, sites with intimal thickness greater than 300 μm (level 4) grow increasingly frequent, creating ample numbers of microregions for evaluation at those excessive depths. These newly created microregions at depths greater than 300 μm acquire ORO staining intensities which quickly come to exceed those found at younger ages (Figures 10-1&2). From age 40 to 70 years, these new depths show rapidly increasing ORO staining; moreover, the shallower microregions at the thickened sites also begin to show increasing intensity of stain with aging. The behavior of the shallower regions is striking: intensities remain nearly level from ages 15 to 35 years when intima is not yet much thickened, but then abruptly change course with further aging after deeper zones of intima have appeared beneath them.

(Technical note: Around ages 35-40 years, region "a" in the graph, Figure 10-2 displays a peculiar pattern that may hold some importance: With the first appearance

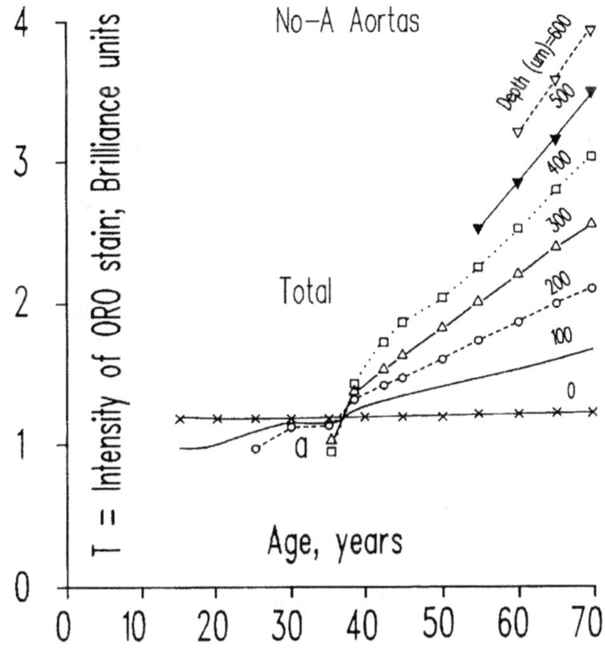

Figure 10-1. Equation (10-2) is plotted with Y = 0. Intensities of total staining predicted from the equation with various values for age and depth into the intima are shown. The position representing intensity 1.0 Bu at age 32 years is marked "a" to aid discussion. Symbols upon the lines are not plotted data points, but only serve to distinguish the lines from each other. Reproduced from reference [64] with permission from Springer-Verlag.

of thickenings sufficient to provide observations at levels 3, 4, and 5 (200-500 µm of thickness), extracellular lipid is generally inconspicuous at those newly emerging levels. The staining intensities even seem to trail what prevails near the intimal surface. The sequestration of lipid in these newly formed masses of fibroplastic tissue seems to require a delay of 5 to 10 years before commencing to accelerate thereafter. It is as if the excessive fibroplasia (F_C in earlier chapters) must pass through an incubation period, analogous perhaps to the rusting of iron, before going into action.)

10.8. ORO stains in relation to age and intimal depth in YesA aortas

The aging YesA aortas closely resemble the NoA specimens in the behavior of *extracellular* lipid at varying intimal depths (compare Figure 10-3 with Figure 10-2).

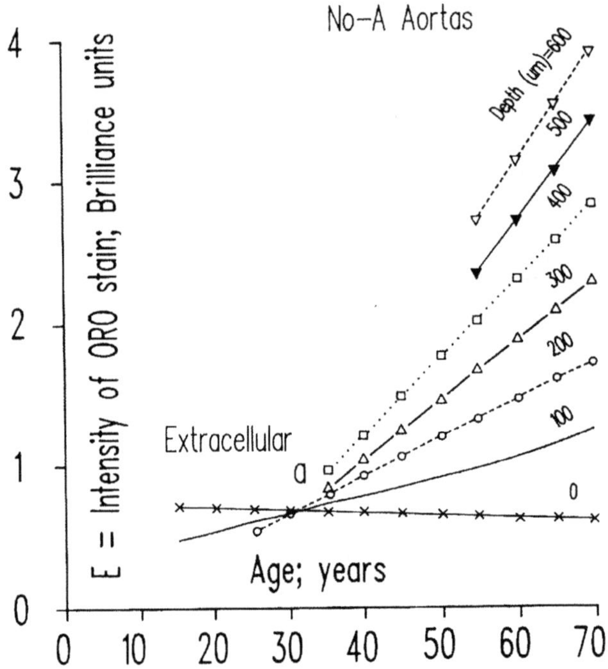

Figure 10-2. Equation (10-6) is plotted with Y = 0. Intensities of extracellular staining predicted from the equation with various substituted values for age and depth into the intima are shown. The position representing intensity 1.0 Bu at age 32 years is marked "a" to aid discussion. Symbols upon the lines are not plotted data points, but only serve to distinguish the lines from each other. Reproduced from reference [64] with permission from Springer-Verlag.

The YesA specimens are remarkable, however, for *total* staining: the subsurface intimal staining, but not the deep staining, greatly exceeds that of NoA specimens, and the difference begins at the earliest appearance of atheromas in the thoracic aorta around ages 35-40 years. The implication is for much greater amounts of *intracellular* lipid in the superficial intima of YesA than of NoA specimens. (This observation is a direct demonstration of the principles embodied in Equation 4-7, Chapter Four, where the YesA condition emerged as a function of both fibroplasia, F_C, and fatty streak elements, P_F).

10.9. ORO stains in relation to intimal thickness and depth

In cross tabulations, the patterns of staining found at increasing depths into intima of

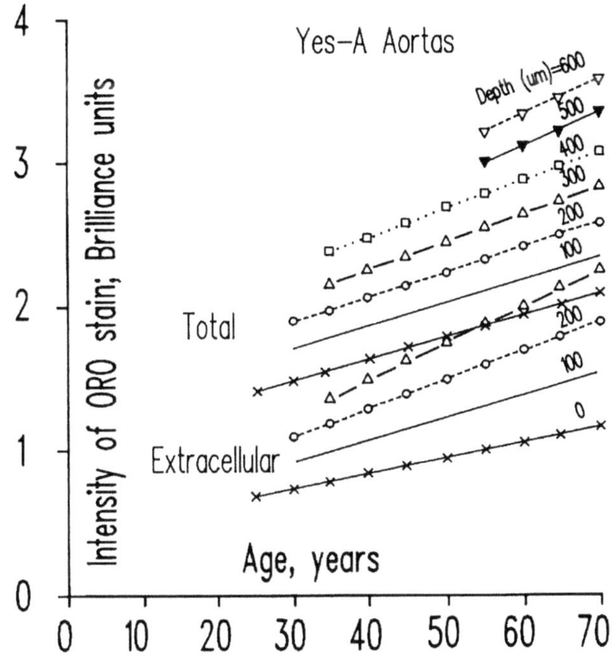

Figure 10-3. Equations 10-2&6 are plotted with Y = 1. Scales are retained for easy comparison with Figures 10-1&2. Reproduced from reference [64] with permission from Springer-Verlag

varying thickness are complex. The intracellular staining generally varies inversely with the extracellular staining, but not always so. Tables 10-5&6 introduce these results by tabulating the ordinates calculated from Equations 10-3&7 (Table 10-4), which omit terms in Y thereby disregarding the YesA status of each case. Table 10-7 is obtained by subtracting entries in Table 10-6 from those in Table 10-5, thereby offering a look at the intensity of intracellular staining. Some prominent features of the patterns are these: (1) At any particular depth and age, the extracellular stain tends to remain constant or decline slightly with increasing intimal thickness (within rows of Table 10-6) while intracellular stain increases with thickness (within rows of Table 10-7). (2) At any particular thickness, intracellular stain declines rapidly with depth (within columns of Table 10-7) while the extracellular behavior depends upon age (within columns of Table 10-6). At age 30 years, extracellular stain is nearly constant at all depths and thicknesses, whereas at age 70 years it rises rapidly with depth. (3) At any particular depth and thickness, intracellular lipid declines with age while extracellular lipid rises.

Table 10-5. Intensity of Total Oil Red O staining, T_C, in frozen sections of lateral thoracic aorta according to depth into the intima by intimal thickness; ordinates computed from Equation 10-3, in Brilliance Units.

Depth into intima, D; N-levels	Thickness of intima, θ; N–levels*								
	1	2	3	4	5	6	7	8	9
			--- AGE = 30 years ---						
1	1.14	1.28	1.43	1.57					
2		0.89	1.11	1.32					
3			0.79	1.08					
4				0.83					
			--- AGE = 50 years ---						
1	1.33	1.49	1.64	1.80	1.95				
2		1.51	1.68	1.85	2.03				
3			1.73	1.91	2.10				
4				1.97	2.18				
5					2.25				
			--- AGE = 70 years ---						
1	1.52	1.69	1.85	2.02	2.19	2.36	2.53	2.69	2.86
2		2.13	2.26	2.39	2.51	2.64	2.77	2.90	3.02
3			2.66	2.75	2.84	2.92	3.01	3.10	3.18
4				3.11	3.16	3.21	3.25	3.30	3.35
5					3.48	3.49	3.49	3.50	3.51
6						3.77	3.74	3.70	3.67
7							3.98	3.90	3.83
8								4.11	3.99
9									4.15

* At ages 25-34 years, thicknesses greater than 399 μm (4 levels)are too few to offer meaningful data; at ages 45-54 years, thicknesses greater than 499 μm (5 levels) are rare.

10.10. Using the sequestration hypothesis to explain Tables 10-6&7

Chapter Eight tells us that the average of fibroplasia per SMC, F_C, is alike at all depths into the aortic intima of any particular thickness. Indeed, the data relating F_C (as the inverse, cells per thickness) to age and thickness are available in Figure 8-3. The entries

Table 10-6. Intensity of Extracellular Oil Red O staining, E_C, in frozen sections of lateral thoracic aorta according to depth into the intima by intimal thickness; ordinates computed from Equation 10-7, in Brilliance Units.

Depth into intima, D; N-levels	Thickness of intima, θ; N–levels								
	1	2	3	4	5	6	7	8	9
			--- AGE = 30 years ---						
1	0.76	0.70	0.64	0.59					
2		0.62	0.61	0.61					
3			0.59	0.63					
4				0.66					
			--- AGE = 50 years ---						
1	1.01	1.00	0.99	0.98	0.97				
2		1.24	1.24	1.24	1.24				
3			1.49	1.51	1.52				
4				1.77	1.79				
5					2.07				
			--- AGE = 70 years ---						
1	1.27	1.30	1.33	1.37	1.40	1.44	1.47	1.51	1.54
2		1.86	1.87	1.87	1.88	1.88	1.89	1.89	1.90
3			2.40	2.38	2.35	2.33	2.30	2.28	2.25
4				2.88	2.86	2.77	2.72	2.66	2.61
5					3.30	3.22	3.13	3.05	2.97
6						3.66	3.55	3.44	3.32
7							3.96	3.82	3.68
8								4.21	4.04
9									4.39

in Figure 8-3are here transferred into Table 10-7 (after inversion to thickness per cell) for ease of further reference. In those tabulations, F_C rises rapidly with intimal thickness and slowly with aging at any particular thickness; but F_C is taken as constant at all depths when matched for age and thickness.

According to the sequestration hypothesis, the magnitude of F_C is thought to govern the increased trapping of extracellular lipids as intima thickens with age. Yet the principle seems to fail when comparing levels of varying depth within sites of anyparticular thickness. Lipid deposition, E_C, increases rapidly with depth even though F_C is constant. Clearly, the sequestering action of F_C is strongly influenced by some

Table 10-7. Intensity of Intracellular Oil Red O staining, I_C, in frozen sections of lateral thoracic aorta according to intimal thickness by depth into the intima; ordinates computed by subtracting entries in Table 10-6 from those in Table 10-5.

Depth into intima, D; N-levels	Thickness of intima, θ, N-levels: 1	2	3	4	5	6	7	8	9
			--- AGE = 30 years ---						
1	0.38	0.58	0.79	0.98					
2		0.27	0.50	0.71					
3			0.20	0.45					
4				0.17					
			--- AGE = 50 years ---						
1	0.32	0.49	0.65	0.82	0.83				
2		0.27	0.44	0.61	0.79				
3			0.24	0.40	0.58				
4				0.20	0.39				
5					0.16				
			--- AGE = 70 years ---						
1	0.25	0.39	0.52	0.65	0.79	0.92	1.06	1.18	1.32
2		0.27	0.39	0.52	0.63	0.76	0.88	1.01	1.12
3			0.26	0.37	0.49	0.59	0.71	0.82	0.93
4				0.23	0.30	0.44	0.53	0.64	0.83
5					0.18	0.27	0.36	0.45	0.54
6						0.11	0.19	0.26	0.35
7							0.02	0.08	0.15
8								-0.10	-0.05
9									-0.24

Age; years	F_C = Fibroplasia per SMC; μm/SMC; applies equally at all depths								
30	4.3	6.6	8.1	9.1					
50	4.4	7.6	10.0	11.2	12.5				
70	4.6	7.7	11.1	12.5	15.9	20.8	23.8	26.3	29.4

additional factor that is governed by distance from the blood stream.

Inspection of Table 10-7 indicates that intracellular staining varies with depth in an approximately inverse correlation with extracellular stain (see note below). In the sequestration hypothesis, extracellular lipid is viewed as a transient inhabitant of

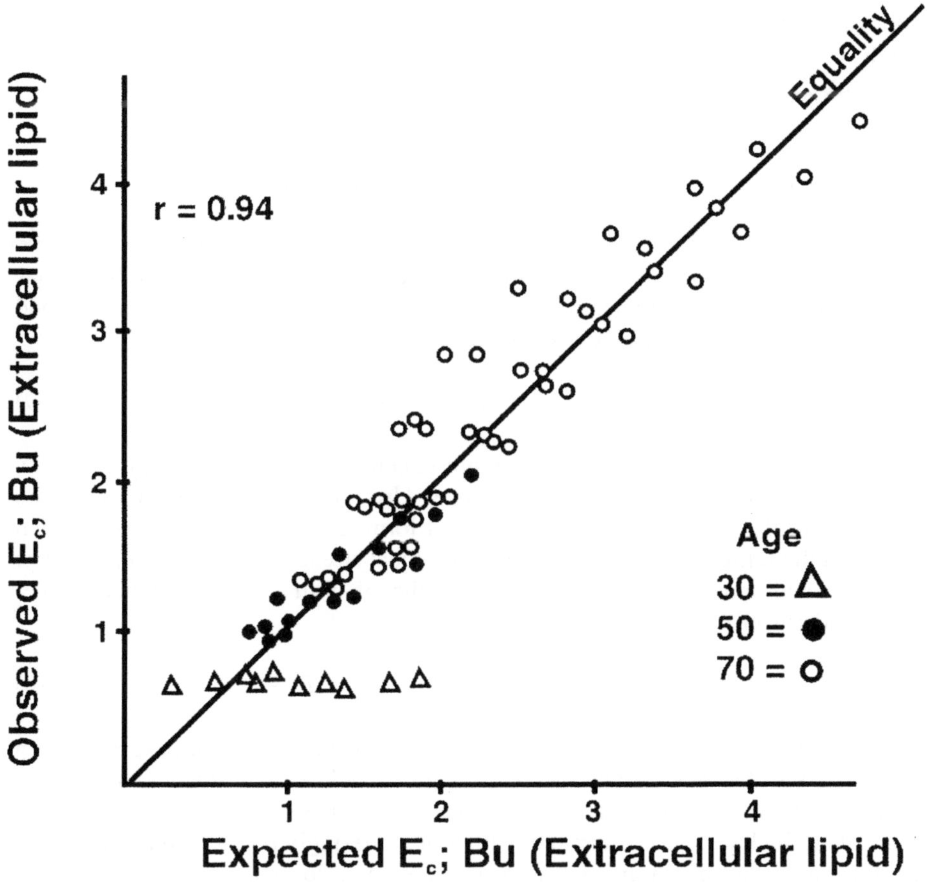

Figure 10-4. Observed values of E_C taken from Table 10-6 are plotted against values expected from Equation 10-9. The diagonal line marked "equality" is a hypothetical entity to be tested against the data; it is not a regression line and is not empirically fit to these data.

fibroplastic intima, to be consumed later by masses of foam cells drawn to this chemoattractant. But foam cells appear to have great difficulty penetrating any great distance from the blood stream. Many reasons might be envisioned as causes for this difficulty, and these reasons are not of immediate concern here. What arises now is that the foam cells laden with intracellular lipid may serve in some way to modify the strengthening affinities for extracellular lipid shown at deep levels. Whereas F_C retains a constant magnitude, the affinity for lipid may be diminished by the presence of foam cells. This matter is taken up next.

(Note: The correlation of I_C with E_C between levels of depth (at each degree of thickness) is not precisely inverse, because the plotted lines in Figures 10-1,2&3 are slightly curved, so that a linear approximation has a nonzero intercept in a later equation.)

10.11. Foam cells as agents for clearing sequestered extracellular lipid

The entries in Tables 10-6&7 can be treated as a body of data to analyze by multiple regression. We wish to explain the occurrence of extracellular lipid, E_C, by the actions of fibroplasia, F_C, and of another influence that diminishes with distance from the blood stream and which we can monitor by measuring I_C.

$$E_C = 0.104F_C - 1.96I_C + 1.22 \quad (R^2 = 0.886) \tag{10-9}$$

By multiple linear regression the above expression rejects thickness, θ, and depth, D, as adding no further statistical significance. Use of standard deviations allows conversion of the equation to the standardized form, $E_C = 0.882F_C - 0.646I_C$, which shows that the regressor variables are about of equal strength in determining E_C. Age enters with additional statistical significance, although with a weak coefficient (0.322 when standardized).

$$E_C = 0.077F_C - 1.86I_C + 0.0212A + 0.35 \quad (R^2 = 0.935) \tag{10-10}$$

Figure 10-4 plots the observed values of E_C against the values expected from Equation 10-9. At ages 50 and 70 years, Figure 10-4 tells us that Equation 10-9 successfully predicts approximations to the observed values of E_C at all depths in aortic intima of all thicknesses. The approximations are amazingly precise, considering that age, thickness, and depth are not used as regressor information. The lessening of extracellular staining near the intimal surface is therefore quantitatively commensurate with abundant foam cells in those locations, *and requires no further information for its explanation.* It is a small step from here to suggest that foam cells are themselves the influence that removes extracellular lipid from the superficial levels of sites that are otherwise prepared to sequester lipids.

This result has the salutary consequence that foam cells stand as sentinels to the former presence of shallow extracellular lipid and to the current presence of deeply trapped lipids. The presence of foam cells was extensively used as sentinels in this way in all earlier discussions of the sequestration hypothesis.

(Note in Figure 10-4: At age 30 years, observed E_C is not commensurate with expected E_C when the latter exceeds 1 Bu of intensity (which occurs at the most thickened sites). It is this discrepancy that allows age to enter into Equation 10-10. This matter arose earlier when discussing region "a" of Figure 10-2. It looks as though a lag of 5 to 10 years is needed for the sequestering conditions measured by F_C to ripen into effective action. Thus, it may not be F_C itself that actuates the trapping of lipids, but rather an aged form of the fibroplastic matrix altered progressively over time.)

10.12. Extracellular lipid as an attractant for foam cells

At age 30 years, intracellular lipid can be found at sites of apparently "normal" intima, lacking any sign of fibroplastic arteriosclerosis. On the cover of this book, this feature is illustrated in Frame "1c". Subjects with this feature were discussed in relation to the "false negatives" observed in Figure 4-1 of Chapter Four, in whom atheroma is thought to evolve from this substrate. This route to atheroma is widely thought to be the only possible route, thereby omitting inquiry into the aging risk factor.

At age 50 years, however, events of overriding importance emerge that are rarely seen in youth. Extracellular lipid appears in ever increasing amounts, at increasingly thickened sites, and especially in the depths of those sites. As the intima grows thicker, foam cells increasingly mass at these sites, and the massing is greatest near the blood stream while tapering away into the intimal depths favored by the extracellular lipid. In the following equation, the attraction for foam cells, I_C, by intimal thickness, θ, can be ascribed to a hypothetical attractive force emanating from the lipid sequestered in the depths of the thickened sites.

$$I_C = 0.12\theta - 0.34E_C + 0.53 \quad (R^2 = 0.701) \qquad (10\text{-}11)).$$

The inverse effect for E_C, which carries a minus sign, can be seen as the consuming action of foam cells removing the extracellular lipid in the superficial levels of intima. On the cover of this book, Frames "2c" and "3c" illustrate the commonplace appearances that result from the balance of these countervailing influences at varying distances into the intima and away from the blood stream..

10.13. Emergence of atheroma

YesA aortas had a tendency to display greater amounts of superficial intracellular lipid than NoA specimens matched for age and intimal thickness (compare "Total" lines in Figures 10-2 and 10-3). If sequestered lipid is the attractant that draws masses of foam cells, then that attraction seems greater in the YesA specimens. The amounts of extracellular lipid could not be distinguished between YesA and NoA specimens in similar comparisons, indicating a difference in lipid composition or in the individual response to that lipid (related perhaps to risk factors such as blood pressure or lipoproteins). This observation may have some relevance to another phenomenon of interest:

Foam cells, reflected in I_C, are seldom seen at depths beyond 300 μm from the lumen. After the emergence of the necrotic core in an atheroma, however, foam cells can be found at unlimited depths interfacing with the luminal side of the lipid core [4,8,20,47,64]. These observations generate an impression that foam cells are being drawn to the sites of deep sequestered lipid. These cells usually fail to migrate deeper than 300 μm, but they overcome that resistance when the necrotic core begins to send out signals of a different sort. The lipid laden macrophages at the interface of fibrous cap with lipid core may well be a different sort than the usual fatty streak elements. Whereas the fatty streak elements have all the usual features of chronic round cell inflammation, the atheroma demonstrates the features of granuloma, including epithelioid cells, foreign body type giant cells, and necrosis of a kind that is reminiscent of caseation. The granulomatous process, apparently, can overcome the usual reluctance of macrophages to migrate far from the blood stream.

CHAPTER ELEVEN

LOCAL SEQUESTRATION OF LIPID FROM PLACE TO PLACE WITHIN AN ARTERY

Abstract. A new method for staining lipid in paraffin sections was applied to a selection of 89 subjects aged 31-92 years. From place to place within each specimen of coronary artery, extracellular lipid strongly tended to select intimal microregions with low SMC densities. Direct visualization of lipid by this new method therefore confirms a key feature of the sequestration hypothesis.

Previous chapters utilized mathematical modeling as a way to deduce the expected properties of extracellular lipid deposits in arterial intima. The incrimination of sequestered extracellular lipid as the culprit for mediating the aging risk factor was deduced indirectly without ever observing the putative agent itself. Chapter Ten used fat stained frozen sections to observe the extracellular lipid deposits themselves, and achieved notable success in observing directly some of the predicted results. That methodology, however, has three principal drawbacks for present purposes. (1) The sections offer only limited structural details, in part because the sections must be exceptionally thick; enumerating SMC nuclei is not sufficiently reliable in such preparations. (2) The lipid does not remain in place during sectioning and staining, but rather diffuses unpredictably and tends to aggregate into droplets of varying sizes; the colocalization of lipid with SMC's is therefore in doubt. (3) It is impractical to apply the tedious and costly method to the entire length of a specimen. Chapter Ten therefore sought only to evaluate the averages for lipid deposits in representative samples of specimens, rather than to examine the detailed dispersal from place to place in each specimen. New methodology now allows us to overcome these drawbacks.

This laboratory recently introduced a new method to fix lipids for later demonstration by fat stains in paraffin sections [75]. This method allows direct observation of how extracellular lipid might colocalize with SMC nuclear profiles. This chapter reports results from applying this methodology to a recently assembled subset of the coronary arteries discussed in earlier chapters.

11.1. Summary of methods

Coronary arteries were sampled from 117 of the autopsied subjects of ages 15-92 years assembled during the years 1998-2001. Cause of death was violence in 80 instances and natural in 37; only 6 of these had various cardiovascular diseases, because instances of such conditions were intentionally avoided. Samples of right coronary artery were split longitudinally for simultaneous processing into H&E stained paraffin sections as usual, and into Oil red O stained paraffin sections in which lipid was previously fixed by the LA/Chr (Linoleic acid- Chromic acid) method summarized in Article 11.10.2 at the end of this chapter. Measurements were made as usual of the variables P_F, P_A, and F_C in the H&E sections. In the fat-stained specimens, measurements were made of atheromas

Table 11-1. Descriptive statistics for selected variables, including product moment correlation coefficients; 89 men and women aged 31-92 years.†

Variable			Correlation coefficients					
Symbol	Mean	SD	E	R_F	R_A	P_F	P_A	F_C
A	53.1	15.7	.23*	.10	.36**	.15	.26*	.41**
E	23.2	22.2	.---	.38**	.36**	.23*	.30**	.52**
R_F	15.5	19.2	.37**	.---	.31**	.79**	.21*	.36**
R_A	9.2	14.6	.31**	.30**	.---	.39**	.82**	.61**
P_F	11.6	16.3	.21	.79**	.39**	.---	.36**	.33**
P_A	10.4	16.2	.26**	.20	.81**	.34**	.---	.66**
F_c	9.8	4.2	.48**	.36**	.54**	.30**	.63**	.---

† Variable symbols as detailed in Table 11-2; A = Age in years. Raw correlations in upper right triangle, age adjusted partial correlations in lower left triangle.
* = Differs from zero at $p < 0.05$, ** = $p<0.01$.

Table 11-2. Numbers of cases and means of selected variables by 15-year age group.

Variable			Age; years				
Name	Symbol	Units	15-30	31-44	45-60	61-75	76-92
FAT STAIN							
Extracellular Lipid	E	%	3.1	19.5	21.6	28.4	31.7
Foam cell Collections	R_F	%	3.8	12.7	18.5	15.7	16.6
Atheromas	R_A	%	0.0	3.3	9.5	17.0	16.7
H&E STAIN							
Foam cell Collections	P_F	%	3.3	8.6	12.5	16.7	12.0
Atheromas	P_A	%	0.5	5.6	10.8	16.1	16.8
Fibroplasia Per SMC	F_c	µm/SMC	5.9	7.9	10.4	10.7	13.1
Number of cases			28	35	28	15	11

Table 11-3. Numbers of cases and means of selected variables by extent of extracellular lipid. Ages 31-93 years.†

Variable Symbol	Extracellular lipid group; %				ANOVA E group		ANOVA Age	
	0-5	6-15	16-35	36-95	F	p	F	p
E	1.0	9.8	25.4	56.4				
R_F	7.7	12.0	13.4	29.1	5.7	0.00	0.0	0.91
R_A	4.1	8.9	7.5	16.5	3.2	0.02	7.5	0.00
P_F	6.3	14.6	6.6	19.1	3.5	0.02	0.9	0.35
P_A	3.9	12.5	8.7	16.5	2.6	0.06	3.1	0.08
F_c	7.5	9.5	10.2	12.0	5.1	0.00	9.0	0.00
A	45.0	50.9	57.9	58.3	6.0	0.00		
Number of cases	22	22	23	22				

† Variable symbols as detailed in Table 11-2. Analysis of variance, ANOVA, was performed with the PROC GLM of the SAS software; F and p statistics for age tabulated here refer to age introduced as a continuous covariate by regression.

in terms of percentage of the specimen length affected, R_A, foam cell collections, R_F, were expressed as percentage of nonnecrotic specimen, and deposits holding only extracellular lipid, E, were expressed as the percentage affected in the specimen at risk, i.e. having neither atheromas nor foam cells.

11.2. Descriptive statistics

Table 11-1 offers descriptive statistics for the pool of 89 subjects over the age of 30 years. Table 11-2 gives numbers of cases and other information by 15-year age group. Subjects less than 31 years of age were omitted in the analyses involving atheromas because atheromas are rare at those ages (means of 0 % of arterial length affected by atheromas in the fat stains and 0.5 % in the H&E stains). Race, sex, and cause of death groups, and subjects of all ages, were combined into a pool after finding that those variables were rejected for inclusion in Equation 1, as explained later.

11.3. Correlations of extracellular lipid with selected variables

The 89 subjects over age 30 years were grouped into intervals of extracellular lipid, E. By coincidence, numbers of cases happened to be almost exactly equal in the four groups. All of the examined variables, R_F, R_A, P_F, P_A, and F_C, rose significantly with increasing E (age adjusted, Table 11-3). Specimens with the most extracellular lipid

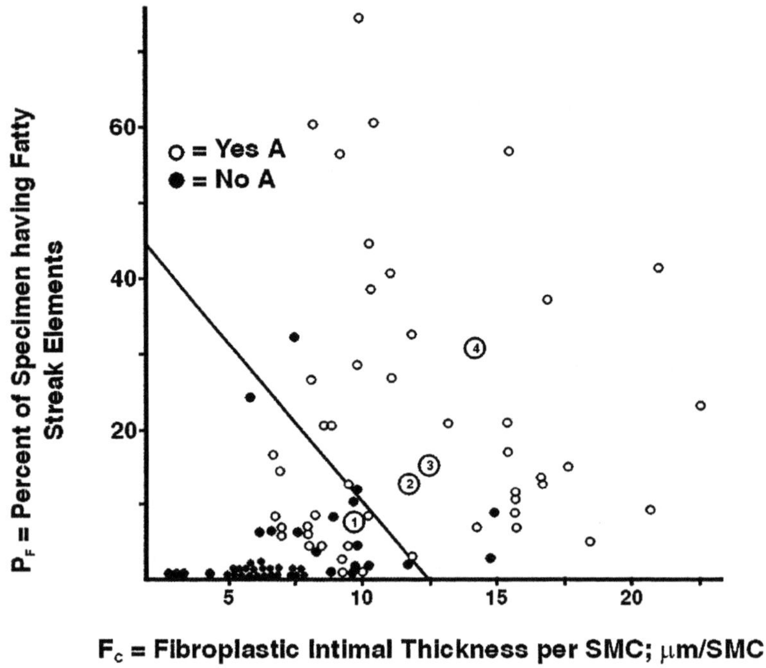

Figure 11-1. Each symbol represents 1 of 89 subjects aged 31-92 years. YesA subjects (circles) have at least one instance of atheroma in the coronary artery specimen while NoA subjects (dots) do not. Sloping line represents Equation 11-1 with W = 0 and this optimally separates circles from dots. Numerals 1, 2, 3, and 4 plot the means from Table 11-3; "4" represents the 22 subjects with the most extensive extracellular lipid deposits.

also tended to be those with the highest values of all five other variables. Nevertheless, in a multivariate setting of Equation 11-1, the correlation of E with F_C vanishes because of multicolinearity effects, as explained next.

11.4. Relationship of atheroma to variables P_F and F_C

In the 89 subjects over 30 years of age, P_F correlates significantly, although weakly, with F_C, r ranging from 0.30 to 0.36 in the various estimates of Table 11-1. The correlation is visible in the scatter plot of Figure 11-1. The scatter plot illustrates the strong tendency of YesA arteries (those with at least one instance of atheroma in the specimen) to display high values of P_F or F_C or both. The sloping line plots Equation 1 with W = 0, which optimally separates the circles (YesA) from the dots (NoA). (Symbols are described in Chapter One, Article 1.7.)

$$W = 0.0779P_F + 0.349F_C - 4.2 \quad (T^2 = 2.92 \quad \%{=}79) \tag{11-1}$$

Standardized coefficients are obtained through multiplication by respective standard deviations to yield $W = 1.27P_F + 1.48F_C$, showing that F_C has about equal weight as P_F in distinguishing atheroma classes. The equation rejects age, race, sex, and all other variables in Table 11-1 as not statistically significant for improving classification of cases. The sloping line correctly classifies 32 of the 47 YesA cases (32 % false negative) and 38 of the 42 NoA cases (10 % false positive). Numerals 1, 2, 3, and 4 in Figure 11-1 plot the means of the four groups in Table 11-3, a plot which shows that subjects high on both P_F and F_C also tend to have much extracellular lipid (group 4). Nevertheless, measurement of E (extracellular lipid) does not help to classify the cases as YesA or NoA. The high values of E occur in subjects with high values of P_F and F_C, and for classification we need only know P_F and F_C (E is rejected as not significant for inclusion in Equation 11-1).

11.5. SMC density at intimal regions with extracellular lipid

Of the 89 coronary arteries from subjects over age 30 years, 12 had no deposits of isolated extracellular lipid in any part of the specimen. In each of these specimens, 9 sites were assessed for SMC densities (density being the inverse of F_C), yielding a total of 108 examined sites. The mean SMC densities were generally around 19 SMCs/(100μm)^2; and it seems appropriate to use this value for the "normal". When each of the 108 sites was divided into the inner, middle, and outer thirds of the intimal thickness, SMC densities prevailed alike at all depths into the intima with no tendency to vary by distance from the luminal surface (Figure 11-2, leftmost triplet of columns). Of the 77 specimens having some isolated extracellular lipid, only 9 were suitably prepared for detailed assessment as follows (most had artifacts of preparation and too few replicate sites for statistical value). These 9 specimens revealed deposits of isolated extracellular lipid in 62 regions. At the lipid infiltrated sites within these 62 regions, SMC densities averaged extraordinarily low values of 7.0 SMC's/(100μm)^2, profoundly below the "normal" value of 19 SMC's/(100μm)^2. The lipid free superficial regions adjacent to the extracellular lipid deposits also averaged low cell densities (7.9 SMC's/(100μm)^2, while the lipid free deep regions between lipid deposits and media held cell densities not significantly different from normal (17.9 SMC's/(100μm)^2 (cf. Figure 11-2, rightmost triplet of columns). Sites distant from the deposits of lipid in these 9 arteries also revealed cell densities not significantly different from normal in the middle third (14.9 SMC's/(100μm)^2 and outer third (17.5 SMC's/(100μm)^2 of intimal thickness. However, those distant sites had significantly depressed cell densities in the inner third of intimal thickness adjacent to the luminal surface (10.1 SMC's/(100μm)^2 (cf. Figure 11-2, middle triplet of columns).

11.6. Photographic illustrations of hypotheses

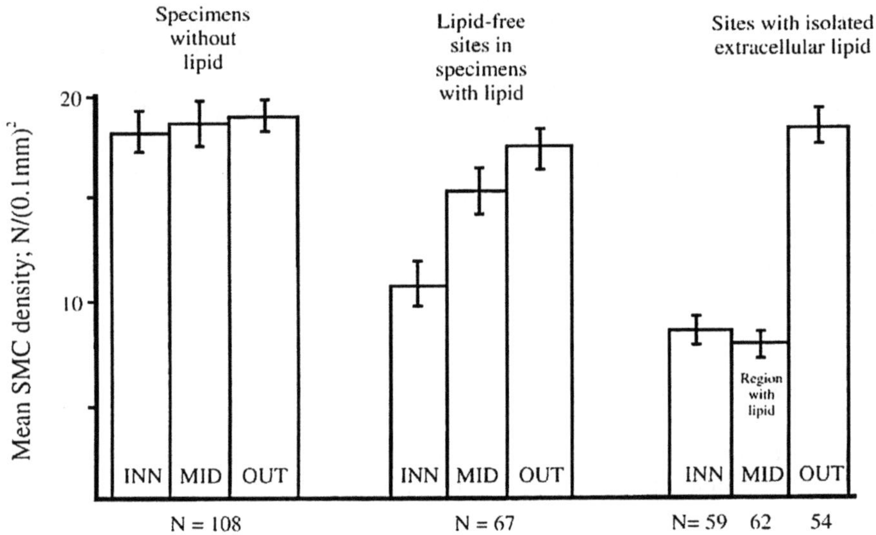

Figure 11-2. The leftmost triplet of columns represent the inner, middle, and outer thirds of intimal thickness in 108 regions (9 sites in each of 12 arteries) observed in specimens that lacked any amount of extracellular lipid. Rightmost triplet of columns represent the 62 sites containing extracellular lipid deposits in 8 suitably prepared specimens observed in the regions with lipid (MID) and in the lipid free superficial (INN) and deep (OUT) regions adjacent to the lipid deposits. The middle triplet of columns represent the inner, middle, and outer thirds of intimal thickness in the 67 lipid free sites within the 8 arteries that did contain some extracellular lipid elsewhere in the specimens. Photograph A in Figure 11-3 exemplifies the leftmost columns, photograph B exemplifies the middle triplet, and photograph C exemplifies the rightmost triple.

The photos of Figure 11-3 can be matched with the triplets of columns plotted in Figure 11-2; the SMC densities visible in the appropriate regions of the photos correspond to the heights of the respective columns. Photograph A illustrates equal SMC densities at all depths into the intima, in keeping with the equal heights of columns in the leftmost triplet of Figure 11-2. Photograph C illustrates diminished SMC density at sites affected by extracellular lipid and in the adjacent lipid-free superficial (but not deep) regions. Specimens having extracellular lipid somewhere in the sample also differ from normal at sites distant from the lipid (Photograph B).

11.7. From sequestered lipid to atheroma

By knowing P_F and F_C we can predict the YesA status of an artery with impressive accuracy, as shown in Figure 11-1. Further knowledge of isolated extracellular

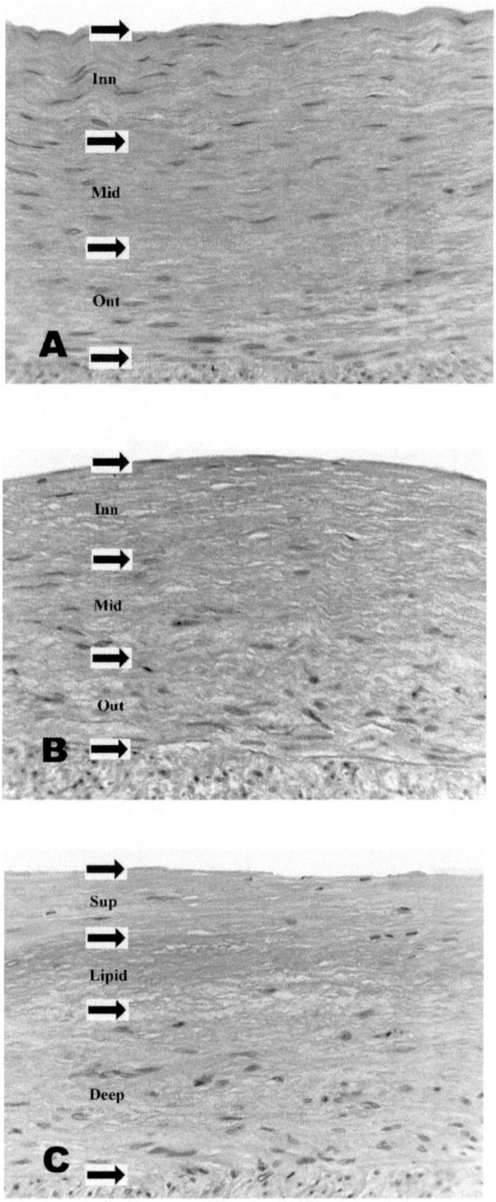

Figure 11-3. Illustrative appearances correspond to the three triplets of columns in Figure 11-2, A to the leftmost, B to the middle, and C to the rightmost triplet. Isolated extracellular lipid (labeled "Lipid" in C), stained red in the original slide, is poorly displayed here in black-and-white, and must be specifically labeled to show its presence in this illustration. (LA/Chr processed paraffin sections, X200.)

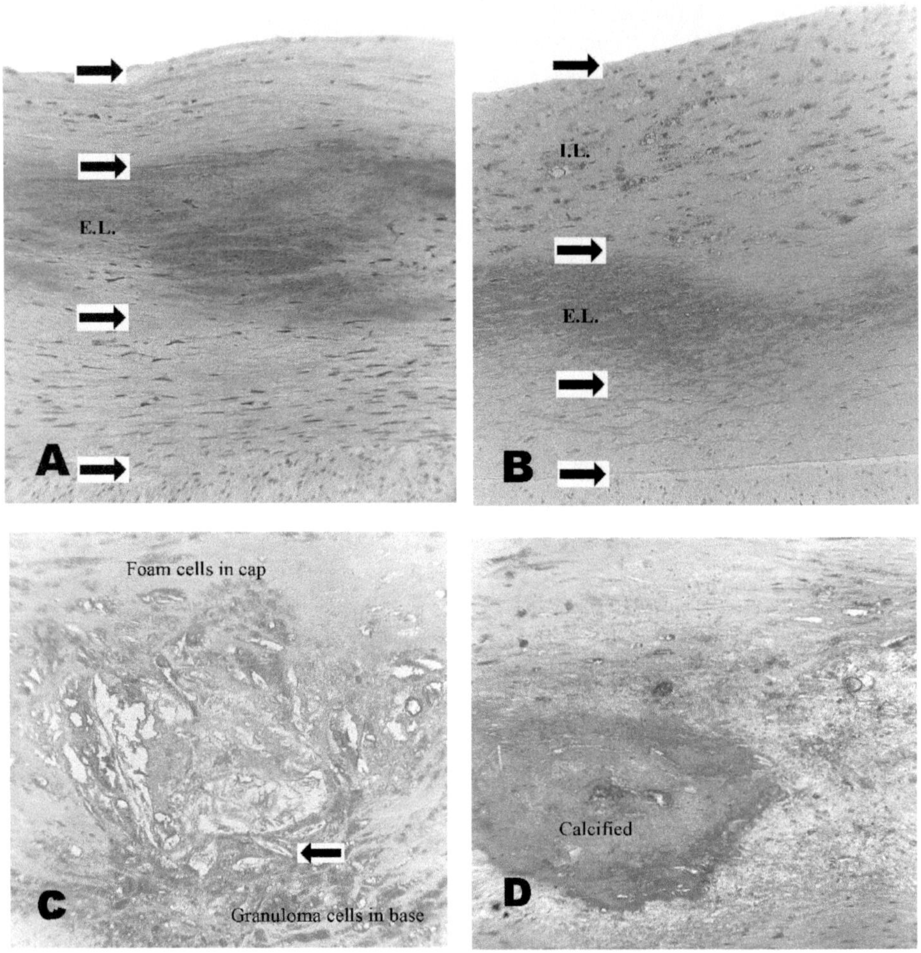

Figure 11-4. Dark zones in A and B between the middle pairs of arrows represent extracellular lipid that was stained red in the origin slides (E.L.); lowest arrows mark intima-media boundaries. Dark region of isolated extracellular lipid in A has no associated fatty streak element, whereas that in B has above it superficial infiltrates of fatty streak elements, foam cells and lymphocytes (I.L). Active atheroma in C has fatty streak elements in the fibrous cap and granulomatous inflamation in the fibrous base; arrow denotes giant cells engulfing impure crystals of cholesterol. Quiescent atheroma in D has lipid debris and cholesterol crystals (needle-like slits) but no fatty streak elements; a node of dystrophic calcification affects part of the necrotic core.

lipid deposits does not help in making this prediction, as if extracellular lipid is simply a coexistent feature also determined by the magnitudes of P_F and F_C. This is the expected outcome predicted by the sequestration hypothesis which proposes that low SMC densities promote atheroma by sequestering extracellular deposits of atherogenic lipids. Unlike P_F, which behaves like a separate process adding its action onto the F_C effect, E behaves like an integral part of the F_C effect with no independent action of its own.

The photographs in Figure 11-4 illustrate the proposed pathogenesis. As aging introduces increasing fibroplasia, thereby diminishing SMC density, this eventually draws extracellular lipid into the prepared sites (Frame A). These lipids change with time, in part from oxidation and glycosylation, until they become toxic so as to exert a chemotactic influence to attract lymphocytes and monocytes that change into foam cells (Frame B). The interaction of foam cells with the toxic lipid then produces an active atheroma marked by granulomatous inflamation with giant cells engulfing toxin-invested cholesterol crystals (Frame C). The atheroma eventually ends its life cycle as a quiescent pool of cleaned and purified cholesterol crystals (the "ashes" of the burned out flame) which no longer incite inflammatory activity, usually calcified (Frame D).

Guyton & Klemp [20] review the evidence from descriptive pathology for this sequence of events, leading them ". . . to pose the hypothesis that tissue growth in the upper intima might occur partly as a pathological reaction to a toxic focus in the deeper intima, perhaps initiated by oxidized lipids . . . We have suggested that extracellular lipid deposits in human arteries may be derived, for the most part, directly from tissue lipoproteins. . . . the earliest manifestation of the atherosclerotic core region is in the deep intima of fatty streaks, well below most of the foam cell lipid." Other authors have also described some of these stages of evolution of the atheroma after its emergence [12,18]. It seems reasonable to suggest that after the atheroma passes through an active phase while disposing of the toxic lipids, it ends in a quiescent phase that is rich in cholesterol, thereby exonerating cholesterol from the list of possible irritants within the necrotic core.

11.8. Speculative synthesis

The outcome subsumed in Equation 11-1, as illustrated graphically in Figure 11-1, applies to comparisons of whole specimens with each other. Figure 11-2 is a more direct test of the sequestration hypothesis. These data compare sites near to each other on a microscopic scale. The sites having extracellular lipid are the ones that display, on average, the lowest SMC densities to be found in the affected arteries, confirming that low densities mark sites that are prone to lipid sequestration. Furthermore, we have reason to think that the low SMC density is a precursor and not a consequence of the lipid deposition. This is because superficial intimal layers that lack lipid nevertheless exhibit low SMC densities in the regions adjacent to the lipid and also at great distances from any lipid deposit.

The data suggest a new line of reasoning: The decline of SMC densities appears to begin near the luminal surface and extend later to the middle levels of intima, seldom ever affecting the deepest levels adjacent to the media. Lipid deposits then tend to avoid the superficial levels, even those of low cell densities, and deposit instead into the middle levels, selecting especially the regions with the lowest SMC densities. The pattern suggests a coexistence of two countervailing influences emanating from the blood stream: One influence promotes fibroplasia, thereby lowering cell densities in preparation for lipid sequestration while the other inhibits the deposition of lipid into the prepared regions. The influence that promotes fibroplasia appears to penetrate more deeply into the intima, with the inhibitory influence restricted only to the most superficial levels.

Certain speculations might flow from this new line of reasoning: The declining SMC densities, sequestration of lipid, and intrusion of atheroma all spare the veins of the body, leading to the proposal that oxygen may the fibrogenic influence emanating from the blood stream. The inhibitory influences penetrate less deeply, becoming sluggish at more that 93 μm from the luminal surface (93 μm being the mean thickness of lipid-free superficial layers in these data). Chapter Ten suggested foam cells as the cleansing influence, which might abandon the site after finishing their job. Another possibility might be something like high density lipoprotein which could do the cleansing by means of its reverse cholesterol transport. Whatever its mechanism, this inhibitory influence offers a possible explanation for the paradox that arteries less than 0.5 mm in diameter never sequester lipid or develop atherosclerosis. The intima of those tiny arteries is, perhaps, never far enough from the inhibitory influence emanating from the blood.

11.9. Conclusions

The right coronary arteries of men and women older than age 30 years often contained deposits of extracellular lipid unaccompanied by atheroma (necrotic core and associated features) or fatty streak elements (foam cells and associated features) (77 of 89 cases = 87 %). These deposits displayed a strong propensity to select regions within each artery that had low SMC densities (7.0 SMC's per $(100\mu m)^2$ of tissue section area compared with 19 SMC's per $(100\mu m)^2$ in normal arteries without extracellular lipid). Patterns in the data support the view that low SMC densities precede and invoke the sequestration of extracellular lipid, in keeping with the predictions from models built in previous chapters.

11.10. Specific methods

11.10.1 Processing of coronary arteries

The proximal right coronary artery was opened longitudinally, cut into three segments of three cm length each, flattened, and fixed in acetate-buffered 4 % formaldehyde. Specimens occasionally required decalcification, using a proprietary hydrochloric acid preparation, RDO (Apex Engineering Products Corp. Plainfield IL). Longitudinally oriented cuts along the myocardial side of the artery were marked with India ink and processed in two ways. One ink marked sample was embedded in paraffin on edge to allow sectioning perpendicular to the luminal surface, and stained with hematoxylin and eosin (H&E). The other facing sample was submitted to lipid fixation before paraffin embedding as detailed below, and sections were later stained with Oil Red O-hematoxylin.

11.10.2 Lipid fixation, the LA/Chr method

As detailed elsewhere [75], blocks of formalin-fixed tissue were immersed in a saturated solution of linoleic acid and lecithin in 70 % ethylene glycol at 56 C; this is done to enrich the lipids in the tissue samples with unsaturated fatty acids. After suitable rinsing in 70 % ethanol and water, the tissues were treated with 2 % chromic acid for 36 hours at 4 C; this is done to fix the unsaturated fatty acids for preservation through paraffin embedding. The final step is a 24 hour exposure to 5 % sodium bicarbonate at ambient temperature; this is done to restore nuclear staining that is weakened by chromation.

11.10.3 Coronary artery assessment

The sections of artery, laid out on the slide in their full length, are scanned and atheromas are marked in black ink on the cover slip; atheromas (AHA Type IV or greater) are usually conspicuous, but in borderline instances can be recognized by finding cholesterol clefts under the 40x objective lens. Foamy macrophages and lipid laden smooth muscle cells, nearly always accompanied by numerous lymphocytes, define “fatty streak elements”; these are marked in green ink. The percentage of the H&E-stained specimen marked in black is called P_A; the percentage of nonnecrotic specimen marked in green is called P_F = 100Green/(1-Black), where Green and Black refer to the respective proportions of the specimen. The same percentages in the ORO-stained specimen are called R_A and R_F respectively. In the ORO-stained specimens, sites with extracellular lipid not already marked black or green are marked with blue ink; the percentage of at-risk specimen marked blue is called E = 100Blue/(1- Black - Green). Specimens having at least one focus of atheroma in the H&E-stained sections are called YesA while all others are NoA. At 9 equally spaced positions along the H&E-stained coronary sample, excluding black and green marked locations and substituting nearby unmarked locations when possible, measurements are taken of intimal thickness, counts are made of SMC’s in a 100 μm wide band through the full intimal thickness, and the ratio of thickness to SMC’s is calculated. Means of these ratios are called F_C.

11.10.4. Comparing sites within specimens

A grid of 1 cm^2 was placed into the eyepiece of the microscope, thereby marking out an area of tissue section of 0.0625 mm^2 when viewed under the 40x objective lens. The grid was placed over every site previously marked blue (containing extracellular lipid) within each specimen. The thickness of the stained region was recorded and the number of associated SMC nuclear profiles falling within the grid were counted. The ratio of cell numbers to intimal area, giving the cell density, is expressed as SMC's per $(100\mu m)^2$. Most of the 76 specimens having some extracellular lipid in them were omitted because of imperfect technical preparation. A total of 62 suitable sites were encountered in 9 cases of near optimal stains. A superficial zone of intima between lumen and lipid was encountered in 59 instances and a deep layer between lipid and media occurred in 54 instances (the lipid strongly tended to seek the mid level and spare the superficial and deep levels). Thicknesses of these layers had mean ± SD of 93 ± 62 μm, 85 ± 47 μm, and 77 ± 42 μm in the superficial, deep, and lipid layers respectively. Cell densities were determined in the superficial and deep lipid-free zones just as in the lipid-containing zones. Sites lacking lipid in these 9 specimens were evaluated by marking each intimal site into layers equaling one-third of the intimal thickness. Cell densities were then determined in the inner, mid, and outer thirds of the intima at each of the 67 such sites that were encountered. In the 12 cases over age 30 years that lacked extracellular lipid, intimal inner, mid, and outer thirds were evaluated for cell density at 9 equally spaced positions along the arteries, giving 108 evaluated positions in all.

CHAPTER TWELVE

FIBROPLASIA IN MICROSCOPIC RENAL ARTERIES

Abstract. Fibroplasia like that in aorta and coronary arteries also occurs in the renal microvasculature. The macroscopic arteries without atheroma, the NoA arteries, have properties quantitatively similar to the microvasculature. The microvasculature never generates the lipid trapping conditions characteristic of macroscopic YesA arteries.

Intimal fibroplasia of the kind seen in the aorta and coronary arteries is also commonplace in the microscopic arteries of various organs. The most affected organs are the brain, kidney, and other solid abdominal viscera. Frequently spared tissues include heart, skeletal muscle, adipose tissue, hollow viscera, and skin. Figure 12-1 compares a typical fibroplastic artery of the renal cortex with the similar features of aorta and coronary artery. In all instances, masses of fibroplastic tissue are molded into intimal structures consisting of dense collagenous layers that harbor longitudinally oriented SMC's scattered individually or in small clusters of SMC's, sometimes in a media-like configuration (cf. Chapter Five).

12.1. Intimal fibroplasia of renal microvessels lacks predisposition to lipid degeneration

In 1911, Allbutt [1] reviewed the knowledge of the time concerning *arterio-sclerosis* and the *arterio-sclerotic kidney*, " . . . by which I mean the state in which this organ, in association with many or most of the other parts of the body, shrinks under a general vascular fibrosis or starvation . . . we must take it for the present from Jores that the arterial change is arterio-sclerotic, as recognized in subintimal hyperplasia with split elastica, and in its disposition to fatty degeneration." Over the decades since Allbutt's era the idea of a general vascular fibrosis has fallen out of favor, even though its name "arteriosclerosis" persists, largely detached from its original meaning.

Figure 12-1 illustrates the kinds of appearances in tissue sections that Allbutt calls, " . . . arterio-sclerosis proper - - that is, subintimal hyperplasia with splitting of the internal elastic layer. Renal arterial disease, then, does not differ from arterio-sclerosis elsewhere . . . a process in nature identical with the process which in large vessels is called atheroma; so that Marchand's and Aschoff's name *athero-sclerosis*, now coming into use, will probably be adopted generally . . . as general arterio-sclerosis penetrates to the finer vessels of the kidney, the case becomes one of renal disease." This statement proclaims atheroma in large vessels as an outcome of the disposition to fatty degeneration in arterio-sclerosis. The wording seems to ascribe a similar disposition upon the microscopic arteries. In truth, the microscopic arteries have no such disposition. If lipid or cholesterol or other features of atheroma are found in these minute vessels, then it is called atheroembolus and judged to be a displaced fragment of a ruptured aortic plaque.

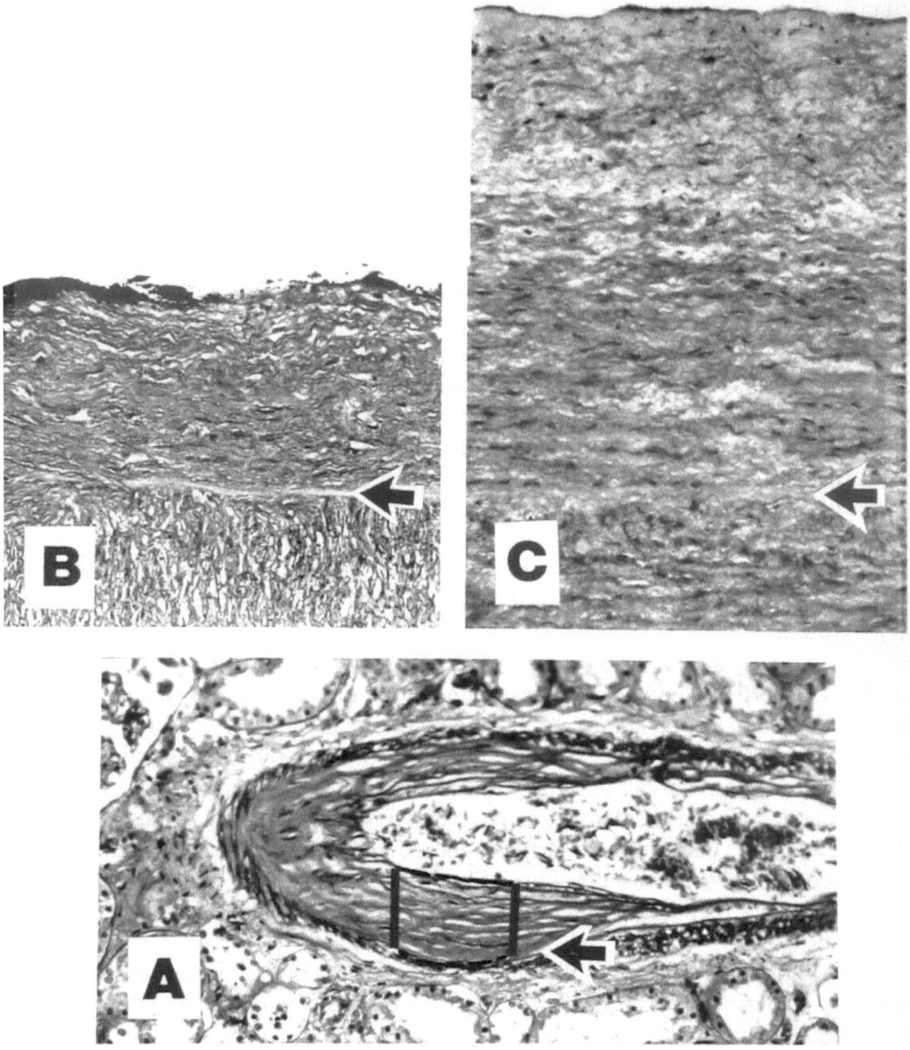

Figure 12-1. Fibroplasia in a renal microvessel (A), coronary artery (B), and aorta (C) are shown as seen in H&E stained paraffin sections. Arrows mark intima-media boundaries. The rectangular area in A spans a width of 100 μm, and the magnification is alike in A, B, and C. Note in A the withering of the media at sites affected by intimal fibroplasia.

The fibroplasia of major arteries is strongly disposed to fatty change (cf. Figures 1-1,2,3, Tables 3-1&3 in Chapters One and Three). In striking contrast, arteries of less than ½ mm diameter exhibit a profound resistance to atheroma or to infiltration by fatty streak elements. Fat stains of frozen sections can reveal some small lipid deposits, usually adherent to degenerating elastica. Stainable lipid rarely forms pools and never draws inflammatory cell responses in the way that atheroma does. The following discussion offers some insight into the profoundly divergent behavior toward lipid performed by the fibroplastic intima of the macroscopic and the microscopic arteries.

12.2. Making of measurements

Observations reported here concern connective-tissue stains of paraffin sections made from perfusion fixed kidneys obtained at forensic autopsies in subjects over 40 years of age (cf. Article 12.10). Figure 12-2 illustrates the kinds of measurements that are made. Elliptic profiles cut obliquely through the cylindrical arteries are identified and the intimal thicknesses and outer diameters are measured at the least axis of each ellipse. The least axis represents the dimensions of the cylindrical artery itself, which are otherwise distorted by oblique planes of cut elsewhere on the elliptic profile. Measurements taken from arteries of various size ranges in a representative pair of kidneys are tabulated in Table 12-1. These tabulations show the kind of skewed bell-shaped curves, as previously encountered for the aorta in Table 2-3 and Figure 5-3 in Chapters Two and Five.

12.3. Meeting the needs of sampling theory

Unlike the lateral thoracic aorta, the microscopic renal arteries do not provide a truly level playing field. Within the broad size ranges used for Table 12-1, some influence by arterial size upon intimal thickness hinders the appropriate pooling of measurements. Moreover, the microscopic arterial tree is built with bifurcations, branches, and bends that all carry some erratic influence upon intimal thicknesses. Nevertheless, we can cautiously extend here the principles of sampling theory developed in Chapter Five in the hope that imperfections in the experimental design will not wholly obscure the useful findings.

12.4. Choosing the realm size, k

Equation 5-4 allows calculation of the realm size, $k = (V_F/M_F)/c$. Use of this equation requires substitute values for the parameter c. Table 5-4 in Chapter Five offers a starting place, because $c = 2.1$ in the aortas with the lowest intimal thicknesses, those of < 50 μm. The intima of renal microvessels averages close to 50 μm in all but the smallest arteries of these five cases, so that $c = 2.1$ seems a reasonable first choice for

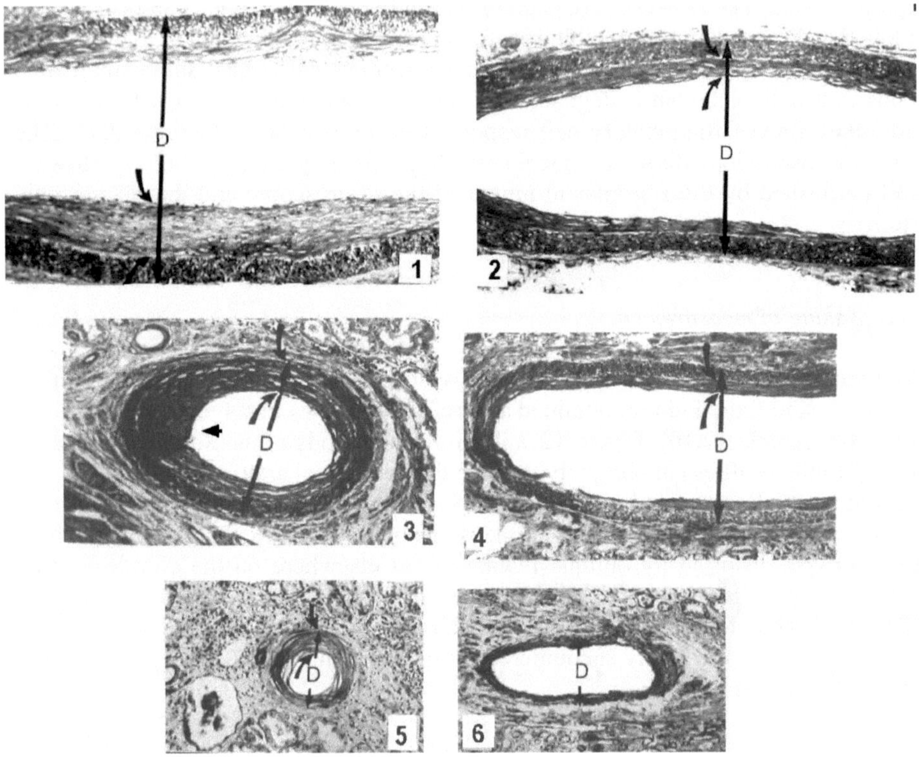

Figure 12-2. Perfusion-fixed specimens stained with PAS are used to measure intimal thicknesses (delimited by curved arrows) within size ranges of arteries grouped by outer diameter, D. Note that measurements must be confined to the least axis of the elliptic profiles, because oblique planes of cut elsewhere on the ellipse distort the measurements (e.g. arrow head in frame 3). The two examples shown here for each size range represent the largest (left) and smallest (right) intimal measurements that were encountered in this specimen.

exploratory beginnings. The observable quantity V_F/M_F, ratio of variance to the mean of intimal thickness, spans the range 4.9 to12.1 in the 30 relevant sets of microvessels. This ratio divided by c is the realm size, k. Hence, a first estimate of k would seem to fall in the range of 4.9/2.1 to 12.1/2.1 or 2.3 to 5.8 μm.

An upper limit on the size of k can be found by looking at the largest value, 12.1 μm, for the ratio V_F/M_F in the microvessels. If we set k equal to this ratio, then this is equivalent to fitting the Poisson distribution, which is the limiting form of the negative binomial when $(V_F/k^2)/(M_F/k) = 1$. This choice is appropriate if we assume the realms to be independent and never aggregated, an unrealistic assumption; hence k

Table 12-1. Numbers of arterial profiles according to thickness of fibroplastic intima in arteries of various sizes in a perfusion-fixed pair of kidneys from a representative forensic autopsy.†

Intimal thickness; µm	Realm N k=5	Range of arterial outer diameter; µm													
		80-99		100-199		200-299		300-399		400-599		600-899		900+	
		O	E	O	E	O	E	O	E	O	E	O	E	O	E
0-5	0	50	60	23	18	1	1	1	0	1	1	0	1	0	1
5.1-10	1	149	131	28	34	5	4	3	2	3	4	4	4	2	2
10.1-15	2	108	108	40	39	5	8	8	4	2	7	4	7	2	3
15.1-20	3	26	39	31	35	9	13	2	7	15	11	13	10	5	4
20.1-25	4	3		31	26	25	17	10	10	25	13	15	12	5	4
25.1-30	5			14	18	20	18	9	11	11	14	15	11	5	4
30.1-35	6			14	11	12	16	14	11	13	14	11	9	1	4
35.1-40	7			10	7	11	13	4	11	5	12	2	7	2	3
40.1-45	8			2	4	4	10	10	10	10	10	4	5	5	3
45.1-50	9			2	2	12	7	8	7	5	8	0	3	1	2
50.1-55	10			1	1	7	5	14	6	8	6	1	2	1	2
55.1-60	11			1	1	2	3	6	5	5	4	4	1	2	1
60.1-65	12					4	2	1	3	4	3	2	1	2	1
65.1-70	13					2	1	1	2	5	2	1		1	1
70.1-75	14						1	4	2	2	1			0	1
75.1-80	15							1	1	1	1			2	
80.1-85	16							1	1		1				
Total N realms		343		197		136		96		115		76		36	
Mean thickness		1.4		3.1		6.3		7.1		6.4		5.1		6.5	
Variance thickness		0.9		4.9		9.5		12.5		11.5		7.4		14.9	
M_1		*		2.4		5.1		5.3		4.7		4.2		4.2	
M_2		*		1.3		1.2		1.3		1.4		1.2		1.6	
Chi Square		*		7.4		13.1		11.8		5.8		7.8		3.2	
(p)		*		(<0.6)		(<0.1)		(<0.2)		(<0.6)		(<0.2)		(<0.7)	

† Chi Square tests comparing the observed (O) with the expected (E) negative binomial fits are calculated after smoothing the data by sliding averages to lessen the problems of rounding errors and digit preference.

*With k = 5, the mean exceeds the variance in this smallest size range, and the negative binomial fit is inappropriate; with k = 3.2, mean equals variance and the Poisson can be fit, and this was done successfully here.

should never be this large. A lower limit on the size of k would seem to reside in the size of the SMC itself. The smallest realms found in the aortic data of Table 5-4 average 2.6 μm, a value that would seem to approach the size of the SMC, and perhaps a modicum of matrix. These considerations place the value of $k = 5$ near the midrange of what seems physically possible for this quantity.

The use of $k = 5$ is mandated by the fact that measurements were all recorded to the nearest 5 μm, thus creating serious rounding-off problems when trying to change the size of the grouping interval. It is reassuring to find the obligatory value of $k = 5$ to be about midway between the upper and lower limits, and also within the range of 2.3 to 5.8 found by direct calculation.

12.5 Fitting the negative binomial

Table 12-1 presents data from a representative autopsy, making use of the choice, $k = 5$ μm. The number of realms, N-realms, range from zero to 16, and the numbers of observed sites with each value for N-realms are tallied. The means and variances of N-realms at the bottom of the table provide the necessary information for fitting the negative binomial distribution to the columns of data. The Chi Square tests fail to reject the fits in the 6 of the 7 instances where this could be tried. The seventh instance was the smallest size range which cannot be appropriately fit this way, because $V_F/M_F = 3.2$ μm, which is smaller than the chosen value of $k = 5$ μm. When the size of k is arbitrarily reduced to permit the curve fitting, the Chi Square test fails to reject the fit in this 7th pool of smallest arteries.

Each of the 6 relevant pools of arteries in the 5 cases, 30 data sets in all, is subjected to Chi Square test of negative binomial fit. Table 12-2 presents the results. Fully 30 % of tests held $p < 0.1$, and 53 % held $p < 0.2$, and this is a significant excess of low p values, implying lack of perfect fit to the data by the negative binomial curves. Many of the observed distributions were overly peaked near the mode with excessive high trailing of the upper tail; 26 of the 30 tests showed the observed upper tail to hold more observations than expected. This kind of discrepancy is anticipated when a pool contains heterogeneous membership, thus generating bimodalities or multimodalities. Possible sources of this type of discrepancy are readily apparent (cf. Article 12.3), and are inescapable imperfections in the experimental design. The magnitudes of these discrepancies are, however, judged to be negligibly small, and to interfere little with subsequent discussion.

12.6. Numbers and cluster sizes of realms

The curve fittings provide estimates for M_1 and M_2 that can serve as parameters of the theoretical curves. M_1 can be visualized as the average number of aggregates and M_2 as the average number of realms per aggregate (by analogy with Equations 5-1&2).

Table 12-2. Numbers of Chi square tests comparing negative binomial forms of distribution to observed data on hypothetical realms by probability levels in 6 size ranges of arteries in 5 specimens of kidney.†

Probability of	Numbers of tests	
Chi square	Observed	Expected
>0.8	5	6
0.5-0.79	4	9
0.2-0.49	5	9
0.1-0.19	7	3
<0.1	9	3

† Chi Square=22.1; $p<0.001$ for agreement of "Observed" with "Expected"

Table 12-3 presents the findings in the 30 data sets. M_2 varies surprisingly little with arterial size. M_2 ranges from the lowest value, averaged over the 5 cases, of 1.10 at sizes 200-299 μm to the highest value of 1.52 at sizes 900+ μm, generally rising with increasing artery size. M_1 averages 3.14 hypothetical SMC clusters in arteries of sizes 100-199 μm, but varies little over the size range 200 - 900+ μm. The mean and the variance of intimal thickness are largely independent of the arterial size over the observed range, so that the parameters derived from them also remain nearly stable.

(Note: Since artery size has little influence on intimal *thickness*, it must therefore alter the intima:diameter *ratios*. Indeed, the ratio displays a peak near the arterial size of 300 μm, declining with increasing or decreasing outer diameter from that peak. The decline into smaller vessels, however, often reverses in hypertension [67]. The 300 μm arteries occur at the corticomedullary junction in the finest arcuate branchings and in the main trunks of the interlobular branches.)

12.7. Visualizing the realm size in tissue sections

Figure 12-1A exhibits a rectangular region of intima in a renal microvessel. This has a width of 100 μm and incorporates the full intimal thickness, by analogy with the similar construction in the coronary artery (Figure 1-1A&B in Chapter One). The rectangle contains nuclear profiles of 9 SMC's within the 70 μm of intimal thickness yielding $F_C = 7.8$ μm/SMC for the fibroplasia per SMC. Applying this methodology to 11 H&E stained specimens from subjects of ages 45 to 82 years, a range of estimates for F_C was 4.9 to 9.1 μm/SMC with an average of 7.6 μm/SMC. (The correlation with age was significant, $r = 0.585$, $p<0.05$.)

The newly found estimates of F_C = 4.9 to 9.1 μm/SMC and the formerly

Table 12-3. Parameters of the negative binomial curves fit to 30 data sets.

Size of Artery; µm	Case I	II	III	IV	V
		--- M_1 ---			
100-199	2.4	2.4	2.7	4.2	4.0
200-299	5.1	5.7	5.9	5.8	6.2
300-399	5.3	7.3	7.6	6.0	7.6
400-599	4.7	7.4	6.6	5.6	6.0
600-899	4.2	7.0	6.8	5.1	6.0
900+	4.2	9.1	6.0	5.6	6.2
		--- M_2 ---			
100-199	1.3	1.2	1.2	1.2	1.0
200-299	1.2	1.0	1.1	1.1	1.1
300-399	1.3	1.1	1.1	1.2	1.1
400-599	1.4	1.1	1.4	1.2	1.4
600-899	1.2	1.4	1.5	1.2	1.3
900+	1.6	1.4	1.6	1.5	1.5

determined values of $k = 2.3$ to 4.8 µm are all aimed at the same physical target, i.e. the size of the space occupied by an average SMC. The choice of $c = 2.1$ was our initial value to begin explorations for determining k. An alternative choice of $c = 1.5$ gives k the range of values, 3.2 to 8.1 µm, as a way to improve its agreement with F_C. The midpoint of F_C is 23 % higher than the midpoint of k, newly assessed with $c = 1.5$. This compares with a similar result for the aorta in Table 5-4, where F_C averages around 19 % higher than k. Again, as for the aorta, the discrepancy vanishes if the width of 100 µm for the rectangle in Figure 12-1A is adjusted to 123 µm.

The new choice for $c = 1.5$ to replace the provisional value of $c = 2.1$ yields estimates of k closer to the practical setting of $k = 5$. The outcome further extends the range of observations that fall consistently into the theory.

Following the development in Chapter 5 we can write, $M_2 * F_C$ as the average thickness of clustered SMC realms. An extreme low for this thickness might be $1.1 * 4.9 = 5.4$ µm and an extreme high might be $1.5 * 9.1 = 13.7$µm. These are predicted values to be tested by future observation.

12.8. A prediction for future testing

The foregoing considerations lead to a prediction of what should be found in future studies of SMC numbers in renal microvessels. If a practical way is found to gather such data, then the ratio of variance to mean of SMC numbers ought to fall near c = 1.5. Furthermore, average thickness of SMC clusters should fall in the range of 5.4 to 13.7 μm. The testing of these predictions would provide a rigorous check on the theory under discussion here. Unfortunately, the necessary measurements are not readily accessible, and may require the use of serial sections.

The measurements of intimal *thickness* exemplified by those in Table 12-1 are attainable because the elliptic profile of an obliquely cut cylindrical artery needs to be assessed only at the site of the least axis of the ellipse. Wall *thicknesses* are geometrically one dimensional, and therefore sufficiently random replicate samples are available in ordinary tissue sections. The *counting of SMC's* is a two dimensional problem that requires an *area* of tissue section. The needs of sampling theory, outlined above, cannot be met, because random replicate *areas* are unavailable.

The 11 kidneys assessed in the earlier discussion were evaluated only for the mean number of SMC's and not for the properties of their dispersal. Even that limited objective is not properly addressed by available methods. Elliptic sections of arteries in tissue sections, used for the counting of SMC's, cannot be chosen at random. The population that is sampled by such counts is ill defined. For this reason, only a token set of 11 kidneys were inspected, and the results are held as tenuous. These results are provisional, and the conclusions are presented only as hypotheses in need of confirmation.

12.9. Lipid trapping and quiescent forms of fibroplasia

The YesA aortas examined in Chapter Five and the YesA coronary arteries of Chapter One reveal certain properties in common. In both instances, the fibroplasia per SMC, F_C, occupied lofty levels in most of the YesA specimens, extending beyond the range of anything seen in NoA specimens. The application of sampling theory now places renal microvessels into this perspective: Even the largest of the hypothetical realms of the SMC found here fall at or below the lowest extreme of YesA macroscopic arteries.

Arteries that harbor instances of atheroma, YesA arteries, were found in earlier chapters to exhibit a peculiar kind of fibroplastic intima, called here by the name fibroplastic arteriosclerosis. That type of intima seems to act as a lipid trap, avidly sequestering atherogenic lipids. The renal microvessels are now found to lack this kind of lipid trapping fibroplasia, and to display instead a kind that powerfully resists lipid deposition. The SMC's in this kind of "quiescent" fibroplasia appear to maintain a proper sort of realm that never expands beyond the harmful threshold. It is this kind of quiescent intima that one finds as a regular feature of NoA arteries.

These considerations lead to a novel proposal: From its youthful beginnings as quiescent fibroplasia, the aging intima seems later to introduce the expansion of SMC realms. This happens early in macroscopic arteries, thus generating conditions for fatty degenerations; it happens later or not at all in the renal microvasculature, thereby forestalling intrusion by lipids. These considerations are added to those in Chapter Eleven, which discusses a possible blood derived influence that prevents lipid deposition within 100 μm of the blood stream.

12.10. Methods

The data reported here are those generated by Clark G Warden, and reported in a thesis delivered to the Department of Biomedical Engineering of Tulane University in partial fulfillment of the requirements of Bachelor of Science in Engineering with Honors, April 14, 1980.

Kidneys were obtained sequentially from a series of forensic autopsies selected only to have ages greater than 40 years. Both kidneys were fixed by perfusion of 4% acetate buffered formaldehyde through the cannulated renal arteries at 100 mm Hg pressure. After 15-30 minutes of perfusion, the organs were then further fixed by immersion in formalin for 48 hours. This treatment leaves the vasculature in a distended state, reversing the usual postmortem collapse of these structures. The fixed organs were cut into 4 mm slices on a meat cutter, in planes perpendicular the general direction of renal artery and vein. All examples of corticomedullary junction exposed in the entirety of both kidneys were submitted for paraffin sectioning and staining with a PAS-Alcian blue procedure [57]. Extensive sampling is required to obtain sufficient numbers of arterial profiles of large size ranges. Each slide was scanned in a systematic fashion grading each artery. A ruler introduced into the microscopic image by camera lucida was used to measure the outer diameter and the fibroplastic intimal thickness at the least axis of elliptic profile of each obliquely cut artery, keep significant figures to the nearest 5 μm.

CHAPTER THIRTEEN

PARAMETERS OF FIBROPLASIA IN RENAL MICROVASCULATURE

Abstract. Intimal fibroplasia in the microscopic renal arteries appears to be of the quiescent form, never emerging in the aggressive form. This conclusion is confirmed here under a wide range of ages and intimal thicknesses. This could explain why atheroma never occurs in this part of the arterial tree, because conditions never exist for sequestration of atherogenic lipids.

Fibroplastic intimal thickenings scattered over the arterial tree have a random component that follows the negative binomial distribution. The mean thickness and the ratio of the variance to the mean determine two parameters that govern the distribution curve. To fit this distribution form to a set of data, it is necessary to group the data points into intervals of increasing thickness. The cutoff points used to group the data serve as a measure of size that converts the continuous variable, thickness, into discrete units. The size to use for this purpose is obtained from biological considerations; it therefore contains information about the biological processes that generate the preatheromatous fibroplasia.

13.1. Source of data

Histologic sections of renal cortex were prepared from samples obtained at forensic autopsies of men and women aged 15 to 79 years, having various causes of death (cf. Article 1.9 in Chapter One). Arteries in the tissue sections were classed into the outer diameter ranges 80-149 μm and 150-300 μm, which are thought to represent resistance level and conduit level vessels respectively. Intimal thickness of each observed artery (call this a "site") is expressed as percent of outer diameter, %od, and these percentages are called R_r and R_c respectively. The manner of making the measurements is illustrated in Chapter Twelve, Figure 12-2.

In this chapter, the %od units are converted to μm of intimal thickness by normalizing to outer diameters of 120 μm and 200 μm through the expressions $1.2R_r$ and $2.0R_c$ respectively. The means of the normalized intimal thicknesses are M_r and M_c and their variances are V_r and V_c respectively.

13.2. The concept of "realm"

Each SMC in the artery wall invests itself with collagenous matrix of varying volume and composition. The space occupied by the matrix of an SMC is called the "realm", which is visualized as a brick shaped volume of space (cf. Chapter Five). We consider here only the thickness of the realm, postponing inquiries of its length and width. The scattering of SMC's therefore implies the scattering of realms, and the distributions of both entities should coincide.

By counting SMC's at 40 sites in the lateral thoracic aorta, we obtain the variance, the mean, and their ratio, V/M = c. The properties of c are explored in Chapter Five and again in Chapter Eight. The value of c ≈ 3.0 was repeatedly encountered under widely varied circumstances. Serious departures from this value could be explained in terms of biased sampling generated through selective censoring by atheromas or foam cell infiltrates, and do not necessarily require biological explanations. The stability of this quantity over varying ages and intimal thicknesses is an especially striking property of the c-parameter, although it can vary greatly from one artery to another.

By measuring intimal thickness at 40 sites in the lateral thoracic aorta, we obtain the variance, the mean, and their ratio, V/M ≈ kc (cf. Equation 5-4). The parameter k specifies the average size of the realms, serving to convert μm of thickness into numbers of realms. Since realms and SMC's coincide, the value of c ≈ 3 can be used with confidence, and k can be determined from the data. The parameter k sets the cutoff points to use for grouping data by intimal thickness, thereby converting the continuous variable into discrete units of size k.

In this chapter we now seek to explore a system where c cannot be measured. Only the product term kc is accessible. This was done in Chapter Twelve after perfusion fixation of arteries in six severely diseased kidneys, with some apparent success. The development here will follow that of Chapter Twelve.

13.3. The size of the realm, k

Variances and means of intimal thickness were used to find the ratios listed in Table 13-1. Those ratios reveal the peculiar property of declining as the intima grows thicker, falling from 8.7 at thicknesses of 0.6 - 4.9 μm to 4.1 at thicknesses of 35.0 - 49.6 μm in the conduit class arteries. This outcome contrasts with the finding of great stability for the ratio at c ≈ 3 reviewed earlier in the aortic data. The ratio measures the product kc, and we can't know whether the decline derives from k or c or both. Yet only narrow ranges for c and k are physically reasonable, and those ranges can be explored here.

The quantity kc can be calculated for each case. By assigning a fixed value to k, we can then calculate c and average the results within groups of cases. Later we shall reverse this procedure.

Chapter Twelve examined the size of k in perfusion fixed microvasculature and settled upon k = 5 as serviceable in a variety of settings. This allows us to write, c = kc/5, where kc is the observed quantity, V/M. The averages of these are given in Table 13-1. By holding k constant, the value of c is seen to fall from 1.8 at thicknesses of 0.6 - 4.9 μm to 0.8 at thicknesses of 35.0 - 49.6 μm in the conduit class of arteries.The c-parameter also declines with age, as seen in Table 13-2. However, this

Table 13-1. Parameters of the negative binomial distribution of fibroplasia in renal microvessels of two size ranges averaged within groups of cases stratified by mean intimal thickness, fixing realm size at k = 5.

Intimal thickness; μm	Var÷ Mean; μm	Number of Realms	c-param	Number of cases
	--- Conduit size; 150-300 μm od ---			
0.6-4.9	8.7	0.3	1.8	37
5.0-9.9	8.7	0.9	1.6	81
10.0-14.9	7.2	1.6	1.5	58
15.0-19.9	6.7	2.6	1.3	63
20.0-24.9	5.8	3.7	1.2	33
25.0-29.9	5.0	5.2	1.1	49
30.0-34.9	4.7	7.0	0.9	16
35.0-49.6	4.1	10.1	0.8	16
	--- Resistance size; 80-149 μm od *---			
5.0-9.9	6.9	1.0	1.7	29
10.0-14.9	4.3	1.8	1.3	31
15.0-19.9	3.3	3.3	1.0	14
20.0-24.9	3.1	6.5	0.7	8

* The groups of cases with < 5 %od intimal thickness is omitted because of difficulties with preference for the digit zero (cf. Article 13.8).

age trend vanishes when comparisons are made within grouping intervals of increasing intimal thickness, because the age trend is merely a reflection of increasing thickness over time (data not shown).

Making use of these values for c allows us to calculate the number of realms of size k = 5. Those numbers range up to 10.1 realms per site in the group of conduit class arteries with greatest intimal thickness. This implies that an exceptionally thickened intimal site should contain about 10.1 SMC's. This seems consistent with appearances such as that in Chapter Twelve, Figure 12-1A. Moreover, fewer SMC's at a site, or more of them, would be inconsistent with visual impressions in unselected specimens, as discussed in Chapter Nine. Hence, k cannot stray far from k = 5, because numbers of realms would then become seriously inconsistent with observed numbers of SMC's, even though the exact enumeration of the SMC's is an unsolved problem.

Table 13-2. Parameters of the negative binomial distribution of fibroplasia in renal microvessels of two size ranges according to age, fixing realm size at k = 5.

Age; Years	Var÷ Mean; µm	Number of Realms	c-param	Number of cases
	--- Conduit size; 150-300 µm od ---			
15-24	9.0	0.9	1.6	45
25-34	9.3	1.5	1.5	67
35-44	7.3	2.4	1.4	92
45-54	5.0	3.5	1.3	78
55-64	5.0	5.0	1.1	28
65-79	4.8	6.7	1.0	24
	--- Resistance size; 80-149 µm od ---			
25-34	8.8	1.6	1.4	4
35-44	9.7	1.6	1.3	9
45-54	6.4	5.7	1.3	28
55-64	5.9	2.1	1.4	17
65-79	10.0	8.0	1.2	20

(Note: The resistance level data are omitted for cases with intimal thickness less than 5 µm because zero readings are so abundant that the relationships become distorted, cf. Article 13.8.)

13.4. Holding the c-parameter to a fixed value

In the aortic data c ≈ 3 was repeatedly encountered. Holding this value constant under nearly all circumstances yielded reasonable estimates for the realm sizes, which expanded with increasing thickness as required by the observed realm size, $F_C = F_1 / C_T$ (cf. Chapter Five). This is not the case for data in Table 13-1. If c = 1.8 found for the group of cases with thinnest intima (where k = 5) is applied to the thickest group, then realm size becomes k = 2.3 µm. If c = 0.8 found for the thickest group (where k = 5) is applied to the thinnest group, then k = 10.9 µm is found. In each of these trials k is found to decline with increasing thickness, exactly opposite of what is expected from findings in the aorta and coronary artery. Hence we obtain reasonable results by holding k constant, but reach absurd conclusions by holding c constant. A variable c is therefore a reasonable deduction, and generates a phenomenon in need of explanation.

Table 13-3. Chi Square tests of negative binomial fits to fibroplasia in renal microvessels of two size ranges pooled within groups stratified into narrow ranges for mean intimal thickness, each group holding 30 cases, and the parameters of the distributions.

Intimal thickness; μm	Number of Realms	Clusters M_1; numbers	M_2; sizes	x where p<x	Number of sites
	--- Conduit size; 150-300 μm od ---				
0.5- 4.2	0.4	0.3	1.3	0.5	292
4.3- 6.5	0.9	0.6	1.4	0.5	253
6.6- 8.4	1.2	0.9	1.4	0.7	274
8.5-10.2	1.5	1.3	1.2	0.7	314
10.3-12.4	1.9	1.6	1.2	0.9	324
12.4-15.1	2.4	2.1	1.1	0.8	280
15.2-17.3	2.9	2.5	1.2	0.2	329
17.4-20.1	3.4	3.0	1.1	1.0	286
20.2-24.8	4.1	3.9	1.0	0.1	259
24.9-27.4	4.8	4.8	1.0	1.0	291
27.5-32.3	5.5	5.7	1.0	0.1	294
32.6-49.7	7.1	7.2	1.0	0.1	309
	--- Resistance size; 80-149 μm od ---				
5.4- 7.4	1.0	0.9	1.2	0.8	285
7.5- 9.4	1.4	1.1	1.3	0.5	286
9.5-12.4	1.7	1.7	1.0	1.0	274
12.7-27.4	2.9	3.1	1.0	0.5	309

13.5. Fitting the negative binomial

SMC's are envisioned as existing in clusters of one, two, three, . . . whose sizes follow a logarithmic distribution with mean M_2. The clusters are then thought to be scattered in accordance with a Poisson distribution with mean M_1. This situation gives rise to the negative binomial distribution for SMC's. The same reasoning applies directly to realms also. We find $M_1 = M_c \log(c)/(c-1)$ and $M_2 = M_c / M_1$ (cf. Equations 5-1 and 5-2), where $kc = V/M$ and $k = 5$.

For fitting of the negative binomial distribution to the data, it is necessary to pool enough cases to obtain sufficient numbers of data points in each pool. For this purpose the cases were ranked on intimal thickness and separated at intervals of 30 cases, thus generating pools of 30 cases of similar mean intimal thickness. A negative binomial curve was fit to the distribution of realms in each pool, and the goodness of fit examined by Chi Square tests. Results of those tests are given in Table 13-3 in terms of the probability levels, $p < x$, where x is tabulated. For instance, the curve fit to the pool with thickness 0.5 - 4.2 μm has $p < 0.5$ meaning that the probability of disagreements occurring by chance alone is less than 50:50 but greater than 40:60. Of the 16 pools, 3 pools obtained $p < 0.1$, when 1.6 pools should obtain this value by chance alone, an insignificant difference. We therefore find no reason to reject the negative binomial as a serviceable description of the data.

13.6. Parameters of the distribution

Increasing intimal thickness from one pool of cases to another in Table 13-3 is attributable entirely to the increasing numbers of realms, M_1.By contrast, M_2 actually declines along with the c-parameter. This would mean, in the context of our theory, that SMC's are initially seeded in clusters, that the SMC's increase in number over time, and that as numbers increase the clusters break up to leave only solitary SMC's at $M_2 = 1.0$. Another way to say this is that we are fitting a Poisson form to the distribution, because that is the limiting form of the negative binomial as $M_2 \rightarrow 1$. We make these deductions because numbers of realms are required to coincide with numbers of SMC's.

It is thought that the neointima under discussion here arises by remodeling of the youthful media of the artery into the neointima. The SMC's that progressively expand the intimal population could be simply relocated medial cells. A precise test of this concept would require quantitative enumeration of medial and intimal SMC's, and the method for doing that is unsolved. Qualitative judgement after the manner of traditional descriptive pathology, however, strongly favors the concept. A close inverse relationship holds between the shriveling media and the expanding intima from site to site in every specimen. Total SMC numbers do not conspicuously change with thickening or aging, as judged by visual inspection. (Note: When SMC numbers visibly increase, this is a strong criterion for the onset of malignant hypertension, a wholly different process extraneous to this discussion.)

13.7. A speculative synthesis

In the coronary artery, a normal intima begins to grow typically around the age of 10 to 15 years, although some sites such as proximal anterior descending branch can experience onset around the time of birth. The initial intima usually resembles media and often cannot be distinguished from media by architecture alone. Later maturation into intimal architecture often leaves scattered media-like islands that appear to persist even to old age (cf. Chapter Nine).

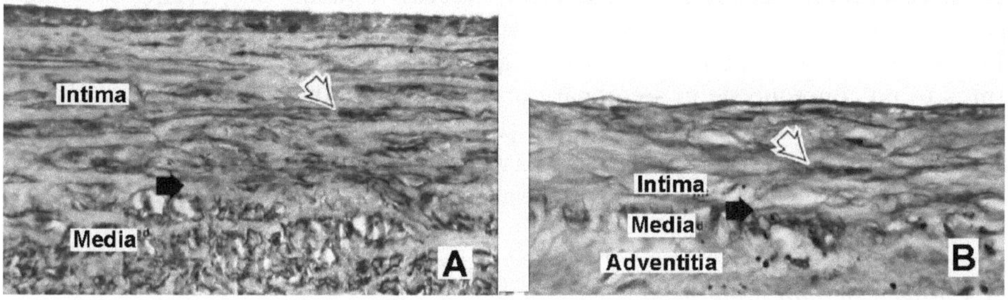

Figure 13-1. A youthful coronary artery, A, and an aged microscopic renal artery, B, have many similarities. Longitudinally oriented SMC's, white arrows, invest themselves with reticulin lamellae that build a layered structure. The layers are about of 12 μm thickness in the coronary artery where each layer contains a cluster of, on average, about two realms, each of 5 μm thickness. The layers in the renal microvasculature are nearer 5 μm, because the SMC's are not clustered. Tissue planes separate artefactually more readily in the micro vessel, giving an illusion of magnified thicknesses. Nigrosin-trichrome, Black arrows mark intima-media boundaries and are of length 20 μm. Reproduced from reference [57] with permission from Springer-Verlag.

We have speculated that those media-like islands could account for the clustering of SMC's that is revealed in the negative binomial curve fitting. If so, the great stability of the $c \approx 3$ in the aortic data, accompanied by a generally stable $M_2 \approx 2$, can be attributed to these residual islands. Now we find a drift toward $M_2 = 1.0$ in the renal microvasculature as intima thickens. Extension of the earlier reasoning encounters the possibility that media-like islands fail to persist in the expanding intima of microscopic renal arteries. Appearances in the microscope seem qualitatively consistent with this conclusion. Perhaps the issue is simply a matter of scale. The sites in the microvasculature are so small that only a few of them would contain an island. Then the sites with islands would simply be counted as heavily populated, and this source of variation would become subsumed in the general Poisson scatter.

All of this leads to a remarkable conclusion: We find in the renal microvasculature no evidence for expansion of the SMC realm with age, as found in the aorta and coronary artery (and also basilar artery [52]). This finding could readily explain the failure of atheroma formation at this level of the vasculature, because the conditions are persistently absent for lipid sequestration. What accounts for this failure, however, remains a mystery without clues.

The findings reviewed here imply that fibroplasia is always of the quiescent form (<9.3 μm) in the microvasculature, and sometimes becomes an aggressive form only in the macrovasculature. This opens yet another puzzle. The rise of blood pressure with age that affects the averages in most human populations can be fully explained by the strictures produced in the microscopic renal arteries by accumulating intimal fibroplasia [58,68-70]. Why should the quiescent form do this, seeing that this quiescent characteristic resembles normal intima rather than aggressive arteriosclerosis (Figure 13-1)? Again, perhaps the issue is a matter of scale. Intimal thickening of 20 μm is of no consequence in a coronary or basilar artery, but this amount produces profound stricture in an artery with 200 μm outer diameter. It seems plausible to suggest that even quiescent arteriosclerosis will suffice to raise blood pressure through the Goldblatt mechanisms of renal microischemia.

13.8. Effects of digit preference

The data fit by negative binomial curves reveal a small but significant digit preference of zero intimal thickness. In all 16 data pools of Table 13-3, the number of sites with zero events consistently outnumbers those with one event by a small margin. For this reason the resistance level vessels with mean intimal thickness less than 5 % are omitted from Tables 13-1&2, because the problem is severe when mean % od = 0 is assigned to a case.

CHAPTER FOURTEEN

THE COURSE OF ARTERIAL INTIMAL FIBROPLASIA IN AGING ARTERIES

Abstract. Arterial intimal fibroplasia, measured by nonatheromatous intimal thickness, increases with age in keeping with a simple formula. The formula applies well to a broad variety of arterial segments and demographic groups of subjects. For each arterial segment, a single variable parameter governs population differences in rates of aging changes. The SMC behavior that propels the aging effect is unknown, but may be a narrow spectrum acting solely upon the formation of collagenous matrix.

Arterial intimal fibroplasia increases progressively with age in the averages of *basal* subjects. The curved lines in Figure 14-1 serve to introduce this topic. The broken lines represent observed data and the solid lines represent the later fitting of the appropriate equation to these data. Line "C", for instance, refers to the renal microvessels in a selected "conduit" size range (150-300 μm, previously discussed in Chapter Thirteen). Within this size range the average ratio of intimal thickness to outer diameter expressed as a percent is called R_c. (Taking the percentage normalizes intimal thickness to outer diameter 100 μm; for some graphic purposes, the normalization is adjusted to od = 150 μm.) The solid curve C summarizes the way R_c progresses with age in this data set, and the gaps between the solid and broken lines serve to measure the disagreements with observed data. Curves A, B, and D offer similar conclusions for grossly visible arteries, aorta, coronary, and basilar, respectively. The curves have sigmoid shapes, with the fastest progression generally around age 40 years and slowing into old age.

14.1. Curve fitting; Step One

The coronary artery and renal microvessels are considered in the following development, because the data sets are large enough to construct subgroups that can be compared for independent confirmation of each other (a kind of "Monte Carlo" approach). This cannot be done with the aorta or basilar artery because too few cases were available. Along with R_c to represent conduit level arteries, another variable, R_r, was similarly constructed for the selected "resistance" level vessels of outer diameters 80-149 μm.

14.2 Step One

The curve fitting begins here with a conventional approach by determining the polynomial regression equations of the form, where y refers to a measure of fibroplasia

$$y = b_1A + b_2A^2 + b_3A^3 + a \qquad (14\text{-}1)$$

in an arterial segment of a *basal* demographic group, and A is age in years. The fitting

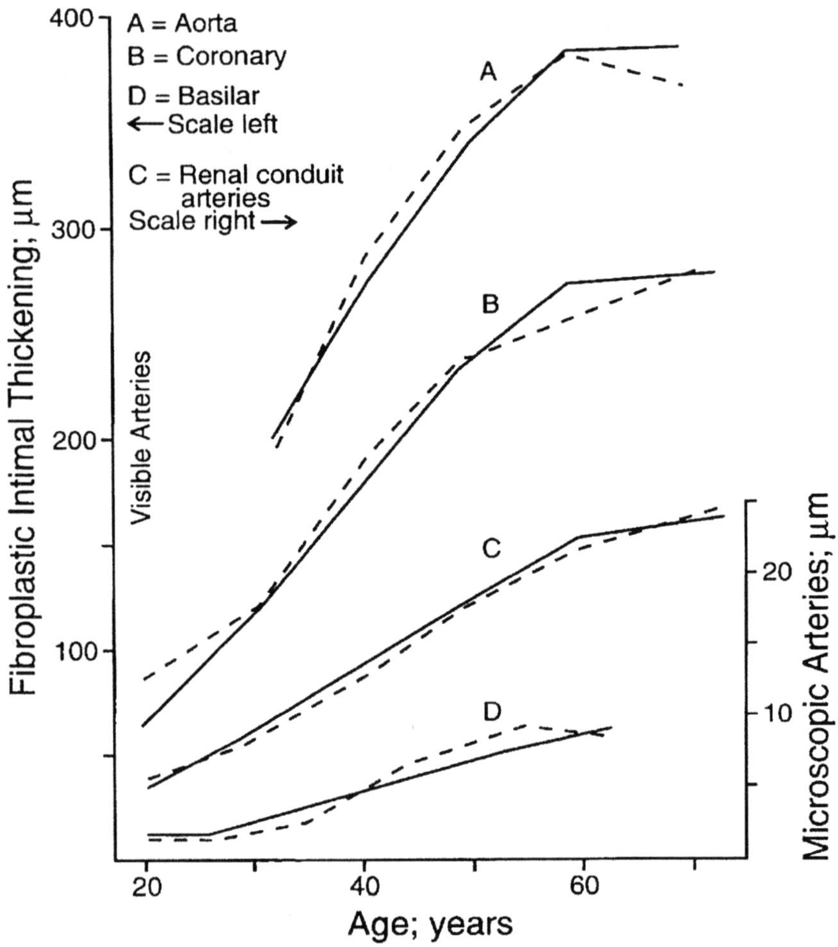

Figure 14-1. Observed data (dashed lines) and solid lines representing Equation 14-2 are plotted. Note that the scale for curve C at the right is expanded from that on the left; in constant units, curve C would overlie curve D. The scale for curve C is normalized to an arterial diameter of 150 µm.

was done over the ages 15-79 years for 12 choices of y, three arterial segments in each of four demographic groups, black and white men and women. In the resulting 12 instances of Equation 14-1, the ratios of the regression coefficients, b_3/b_2, were found to be remarkably constant across demographic groups, and to form a simple pattern of variation across arterial segments, while no such simplicity emerged for b_3/b_1 or b_2/b_1 (Table 14-1). Furthermore, the intercept term, a, was found to be insignificant in seven of the determinations, negative in five, and positive in four, all with small magnitude, and elimination of the intercept produced minimal reductions in R^2 (Table 14-1).

14.3. Step Two

Equation 14-1 was modified by dropping coefficient a, thereby forcing the regressions through the origin. The outcome was, as at Step One, that the ratios b_3/b_2 held close to constancy among the four demographic groups for each arterial segment. Moreover, the coefficient b_1 emerged with little statistical importance; it was rejected outright in seven determinations, and was trivial with inconsistent sign in the other five determinations. This led to the elimination of b_1A and to Step Three, the imposition of the constant ratio, $b_3/b_2 = \alpha$ in a simplified model

$$y = \beta A^2(1 + \alpha A) \tag{14-2}$$

14.4. Step Three

The magnitude of α is estimated by a continuation of the curve fittings in Table 14-1: In the expression $y = b_2A^2 + b_3A^3$, an average of $\alpha = b_3/b_2$ across demographic groups was obtained for each assignment to y. For intima of the coronary, renal conduit and renal resistance arteries, the averages of α were -0.0094, -0.0089 and -0.0044 respectively. (For the coronary artery, $\alpha = -0.01033$ in men and $\alpha = -0.0079$ in women, a statistically significant difference that becomes important in later chapters. For R_c and R_r the sex differences are not significant.)

14.5. Step Four

Consider the Equation

$$y = \beta Z + c_1A + c_2A^2 + c_3A^3 \tag{14-3}$$

where $Z = A^2(1 + \alpha A)$ and α takes the values listed in Article 14.4. In every instance of fitting this equation, the term βZ entered first as statistically strongest, with β taking the values reported in Table 14-2. None of the 36 c_n coefficients subsequently evaluated was statistically significant. These considerations reduce Equation 14-3 back to the form of Equation 14-2 by discarding the insignificant terms. This outcome implies that Equation 14-3 adds no useful information to that of Equation 14-2.

14.6. Parameter β estimation

The parameter β in Equation 14-2 is readily estimated by linear regression, making the change of variable $Z = A^2(1 + \alpha A)$ and fitting $y = \beta Z$ (forced through the origin). This was done here for the fibroplastic intima of the coronary, $\beta = 0.190$, and of the renal conduit arteries, $\beta = 0.00867$, in a pool of all *basal* subjects of ages 15-79 years. These fitted equations are shown as solid curves in Figure 14-1.

For comparison, data from aorta (lateral thoracic segment) and basilar artery were retrieved from the archives of two previously reported studies [52,65]. When data from those two prior studies were taken through the four steps of curve fitting, the end result was Equation 14-2, with $\alpha = -0.012$ and $\beta = 0.24$ for the aorta, and $\alpha = -0.0066$ and $\beta = 0.025$ for the basilar artery. These curves are also added to Figure 14-1.

The plots in Figure 14-1 depict sigmoid curves. In these curves, the locations of the inflection points are at ages 28, 34, and 51 years for the aorta, coronary, and basilar segments respectively, and at ages 38 and 76 years for the conduit and resistance level renal arteries respectively. These inflection points fall at lower ages as α becomes increasingly negative.

14.7. Influence of renovascular fibroplasia on parameter α

The average $R_a = (R_r + R_c)/2$ was calculated for each subject, and, for the pool of all *basal* cases, its median was located in each 10 year age group. The cases above or below the age-specific medians were called High-R_a and Low-R_a respectively.

When data on F, R_c and R_r for these groups were taken through the four steps of curve fitting, the end results were Equation 14-2 in every instance, with values of α persisting as listed under Article 14-4 (development omitted here for brevity).

14.8. Parameter β in High-R_a and Low-R_a subjects

After confirming the values of α to use, High-R_a subjects yielded $\beta = 0.00502$ for R_r, $\beta = 0.01056$ for R_c, and $\beta = 0.208$ for F. Low-R_a subjects yielded $\beta = 0.170$ for F, $\beta = 0.00604$ for R_c, and $\beta = 0.00199$ for R_r.

These six versions of Equation 14-2 are plotted in Figure 14-2. Of the six plots, those for R_r fail to show the expected sigmoid shapes. This failure reflects the fact that $\alpha = -0.0049$ in that setting determines an inflection point at A = 76 years, which falls at the upper end of the observed range of ages. For the graphs, conduit intima is normalized to od 150 μm as in Figure 14-1, while resistance intima is normalized to od 100 μm, by use of the expressions $150R_C$ and $100R_r$. This is done only for graphic purposes and not for fitting of equations.

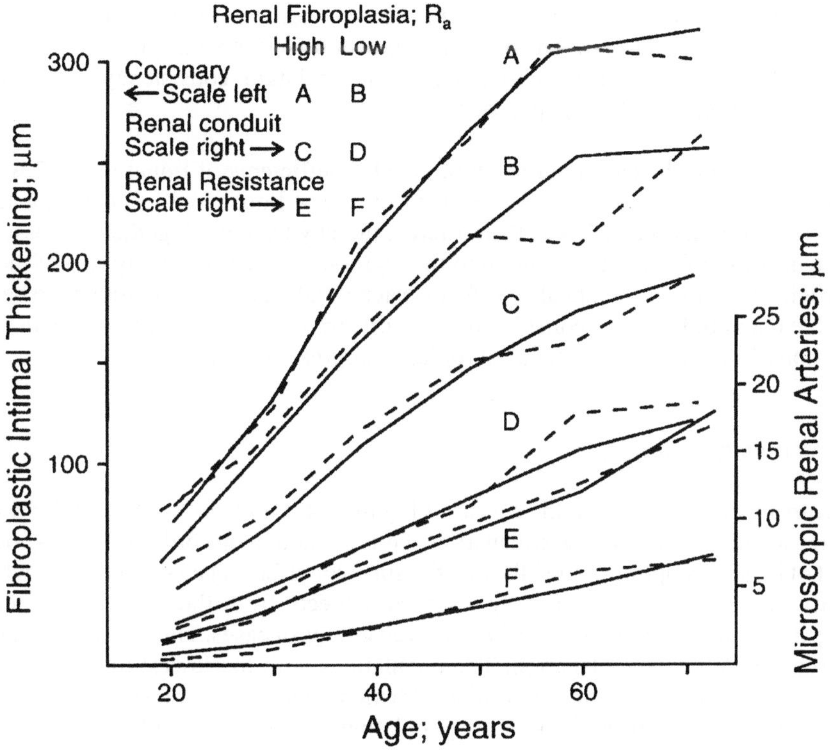

Figure 14-2. As in Figure 14-1. Curves A, C, and E represent subjects above the median of R_a within each 10-year age group, and Curves B, D, and F represent subjects below the median.

14.9. Curve fitting for six other populations

For this analysis, data were retrieved from the archive of a previous study [67]; for comparison, data on New Orleans were copied from Table 14-2. In the previous study coronary arteries were obtained from autopsies in each of the cities listed in Table 14-3. The previous study prepared formalin fixed specimens, longitudinally opened, flattened, Sudan stained, and stored in plastic bags. The study then measured fibroplastic intimal thicknesses in H&E-stained paraffin sections of subjects aged 15-69 years. The *basal* cases from those populations were taken through the four steps of curve fitting.

At the third step of the curve fitting, the magnitude of α was obtained from the ratio of coefficients $b_3/b_2 = \alpha$ as usual. The values of α (Table 14-3) were, with one exception, found to fall within the narrow range of -0.00809 (Sao Paulo men) to -0.01394 (Manila men). This range includes the previously identified values in New Orleans. The mean of these estimates is -0.01068 compared with the value of -0.0094 used in Figures 14-1 and 14-2. The previous value of $\alpha = -0.0094$ is retained in further analyses of these six additional populations to maintain consistency and allow comparisons among all of the populations.

At the fourth step of curve fitting, the β coefficients of Equation 14-2 were obtained and tabulated (Table 14-3). In each instance of Equation 14-2, the expanded list of terms in Equation 14-3 was then offered, thereby testing altogether 48 possible coefficients of the form c_n for addition to the regression equations. Only two of these c_n coefficients acquired statistical significance at $p < 0.05$, and all were judged to be of negligible magnitude. These considerations again arrived at Equation 14-2 as a useful summation of how coronary intimal fibroplasia progresses with age.

14.10. Slackening in old age

One recurring property seen in the curves of Figures 14-1 and 14-2 is the trend toward flattening in old age. A possible explanation of this phenomenon could be a well known artefact of biased sampling, called the healthy survivor effect. The artefact arises when mortality removes the most severely diseased subjects from the older age groups, leaving the healthiest subjects to fill the observed groups of cases. The findings in Table 14-3 argue against this explanation. The populations with the lowest mortality from atherosclerosis, such as Durban Bantu and Mexico City, reveal the same downward curvature in old age as the high mortality populations, such as New Orleans blacks and whites (this is confirmed because a stable value of α across populations cannot be rejected statistically). It is hard to invoke the healthy survivor effect in Durban or Mexico City, because the mortality rates from atherosclerosis are low enough to exert only limited depletion of the aged populations.

14.11. Allbutt's view

As noted in Chapter One, Allbutt proposed that arteriosclerosis is a common process, an entity, that falls alike, whether harshly or mildly, on the arteries throughout the body. Figure 14-2 offers support for this view, because subjects in the high range for fibroplasia in the renal microvasculature also display a high range of involvement in the coronary arteries.

14.12. The tracking of cohorts

An individual is placed into a Low R_a or High R_a group in relationship to his peers. If the variable R_a shows "tracking" from youth to old age, which is to say that the individual tends to remain in the same group throughout life, then the High R_a groups in all of the 10-year age ranges represent cohorts with the similar trait of persistently exceeding the average level of arterial fibroplasia. It is as if following cohorts through life insofar as their R_a status is concerned. We have no reason to suspect that subjects will shift from low to high or high to low groups at any time throughout the aging process; if we assume that they do not then this will simplify further discourse.

14.13. Sources of subjects

Samples of coronary artery and renal microvessels are those described in Chapter One, Article 1.9. Cases supplying data on aortas are those described in Chapter Four. Data on basilar artery are retrieved from a previous study [52]. Coronary data on six populations were retrieved from the archive of the International Atherosclerosis Project [67]. The 272 *basal* specimens successfully processed for microscopy were chosen to be representative of 10-year age groups in each population.

Table 14-1. Regression coefficients for various polynomial equations.

		Step One; a included			Step Two, No Intercept			R^2 with b_1A included or excluded			
								Step One; b_1A		Step Two; b_1A	
		b_2/b_1	b_3/b_1	b_3/b_2	b_1	b_2	b_3	Included	Excluded	Included	Excluded
$y=R_r$;	Black female	0.071	0.00023	-0.00461	*	+	-	0.6308	0.6260	0.6810	0.6793
Renal	White female	0.100	0.00022	-0.00417	*	+	-	0.4479	0.4129	0.7029	0.7017
resistance	Black male	0.208	0.00029	-0.00415	-	+	-	0.5899	0.5793	0.7836	0.7836
arteries	White male	0.472	0.00186	-0.00490	*	+	*	0.5020	0.4919	0.7460	0.7446
$y=R_c$;	Black female	0.0061	0.000046	-0.01010	+	*	*	0.7230	0.7229	0.8930	0.8916
Renal	White female	0.0035	0.000025	-0.00813	*	+	-	0.5544	0.5527	0.8759	0.8730
conduit	Black male	0.0064	0.000039	-0.00885	*	+	-	0.6271	0.6220	0.8764	0.8763
arteries	White male	0.0029	0.000217	-0.00862	+	*	*	0.4989	0.4843	0.8373	0.8354
$y=C$;	Black female	0.00403	0.0000334	-0.01064	+	*	*	0.4214	0.4182	0.8176	0.8154
right	White female	0.00068	0.0000857	-0.00840	+	*	*	0.4059	0.3947	0.8513	0.8503
coronary	Black male	0.00072	-0.0000018	-0.00962	*	+	-	0.4037	0.3908	0.7964	0.7953
artery	White male	-0.00610	-0.0000729	-0.01099	*	+	-	0.2874	0.2874	0.7937	0.7929

At Step One of curve fitting, the regression equation, $y = b_1A + b_2A^2 + b_3A^3 + a$, was tested, and the ratios of the resulting b coefficients are shown here. At Step Two, the intercept was forced through the origin; the statistically significant coefficients carried the algebraic signs noted here as "+" or "-", while those not of significance are noted by "*" ($p < 0.05$). Deleting the term b_1A was then attempted at Step One and again at Step Two. The R^2 values obtained in these four fittings are tabulated here. After these two steps, the model carried into the third step becomes $y = \beta A^2(1 + \alpha A)$, which forces a constant ratio of $b_3/b_2 = \alpha$ in all subsequent models. Numbers of cases are Black females = 71, White females = 60, Black males =145, White males = 109.

Table 14-2. Regression coefficients for the equation $y = \beta Z = \beta A^2 (1 + \alpha A)$, estimates of b_3/b_2, and algebraic signs of significant c coefficients in the equation $y = \beta Z + c_1 A + c_2 A^2 + c_3 A^3$ in selected data sets.

		β coeff	R^2 Simple	R^2 Total	c_1	c_2	c_3	b_3/b_2
$y=R_T$	BF	0.00297	0.6780	0.6808	*	*	*	-0.00456
Renal	WF	0.00251	0.6820	0.6940	*	*	*	-0.00611
resistance	BM	0.00369	0.7833	0.7833	*	*	*	-0.00405
arteries	WM	0.00309	0.7431	0.7460	*	*	*	-0.00492
$y=R_c$;	BF	0.00882	0.8916	0.8930	*	*	*	-0.00848
Renal	WF	0.00756	0.8678	0.8689	*	*	*	-0.00856
conduit	BM	0.00861	0.8762	0.8763	*	*	*	-0.00879
arteries	WM	0.00768	0.8356	0.8373	*	*	*	-0.00859
Y=F;	BF	0.120	0.8125	0.8176	*	*	*	-0.00878
right	WF	0.141	0.8632	0.8711	*	*	*	-0.00815
coronary	BM	0.213	0.7931	0.7963	*	*	*	-0.00959
artery	WM	0.234	0.7899	0.7937	*	*	*	-0.01106

The α coefficients were obtained by averaging the four ratios b_3/b_2 derived by the fitting of $y = b_2A^2 + b_3A^3$ in the four demographic groups for each choice of y; these are -0.0094 for F, -0.0086 for R_c, and -0.0049 for R_T. The signs "+", "-", or "*" tabulated for the coefficients c_1, c_2, and c_3, are done as in Table 14-1; all of them encountered in this table lack statistical significance. The "simplified" R^2 refers to the equation $y = \beta Z$, and the "total" R^2 obtains after entry of all terms, significant or not.

Table 14-3. Regression coefficients for the equation $y = \beta Z = \beta A^2 (1 + \alpha A)$ in selected data sets, using fixed values of α, and algebraic signs of significant c coefficients in the equation $y = \beta Z + c_1 A + c_2 A^2 + c_3 A^3$.

Population	Number Of cases	β coefficient	R^2 Simplified	R^2 Total	c_1	c_2	c_3	α when estimated
			Men					
Durban Bantu	23	0.083	0.7126	0.7126	*	*	*	-0.01128
Sao Paulo White	25	0.097	0.7246	0.7246	*	*	*	-0.00809
Mexico City	28	0.101	0.8050	0.8050	*	*	*	-0.01126
Bogota	26	0.129	0.8035	0.8035	*	*	*	-0.00434
Durban Indian	23	0.145	0.8066	0.8066	*	*	*	-0.00906
Manila	21	0.186	0.7756	0.8755	*	*	-	-0.01394
New Orleans Black	145	0.213	0.7931	0.7931	*	*	*	-0.00959
New Orleans White	109	0.234	0.7899	0.7899	*	*	*	-0.01106
			Women					
Durban Bantu	22	0.066	0.7426	0.7426	*	*	*	-0.01197
Sao Paulo White	25	0.082	0.7948	0.7948	*	*	*	-0.01137
Mexico City	29	0.124	0.8469	0.8469	*	*	*	-0.01103
Bogota	13	0.112	0.8350	0.8350	*	*	*	-0.01177
Durban Indian	18	0.150	0.7895	0.7895	*	*	*	-0.01083
Manila	19	0.143	0.8505	0.9125	*	*	-	-0.01323
New Orleans Black	71	0.120	0.8125	0.8125	*	*	*	-0.00828
New Orleans White	60	0.141	0.8632	0.8632	*	*	*	-0.01054

The fixed value of α is $\alpha = -0.0094$ obtained for the pool of New Orleans men and women in Table 14-2. The signs "+", "-", or "*" tabulated for the coefficients c_1, c_2, and c_3, are done as in Table 14-1. The "simplified" R^2 refers to the equation $y = \beta Z$, and the "total" R^2 obtains after entry of all terms, significant or not. When α is estimated as explained in Step Three of curve fitting, it takes the tabulated values; these are not used for calculating the β coefficients tabulated here, which relate to the circumstance of $\alpha = -0.0094$, using a common value for all demographic groups.

CHAPTER FIFTEEN

THE COURSE OF FIBROPLASIA PER SMC OVER TIME IN AGING ARTERIES

Abstract. Total fibroplastic intimal thickness, F, comprises the two independent aspects, the aspect related to thickness per SMC, F_C, and the aspect imparted by numbers of SMC's, C_T. Data on C_T can therefore serve to measure its associated aspect of total intimal fibroplastic thickness, F. The two aspects of fibroplasia are here denoted the F_C-linked component and the C_T-linked component of arterial intimal fibroplasia, where $F = C_T * F_C$.

The development here returns to Chapter Fourteen, sequentially setting $y = C_T$ and $y = F_C$ in Equation 14-1. The steps of curve fitting are the same as before. The detailed tabulations of the model building are omitted here for brevity, because they closely imitate the patterns shown in Tables 14-1,2&3. The coronary arteries of 357 *basal* men and 191 *basal* women, aged 15-79 years, are examined separately because of finding important sex differences. These include the subjects detailed in Chapter One, Article 1.9, with series expanded by additional ongoing case collection.

15.1. Step One, with $y = C_T$

At Step One of curve fitting, the intercept, a, was rejected as before, for the two pools of cases, men and women.

15.2. Step Two, with $y = C_T$

Step Two rejected the cubic term on age, b_3A^3, for both men and women. The reduced equation became $C_T = b_1A + b_2A^2$.

15.3. Step Three, with $y = C_T$

Multiple versions of the equation at Step Two, fit to men and women of the various groupings used in Chapter Fourteen, yielded b_2/b_1 within a narrow range around the average of -0.008675. When fit to the 357 men, the 191 women, and the pool of all 548 subjects, the results were $\gamma = b_2/b_1 = -0.00905, -0.00878, -0.00910$ respectively. This outcome led to the simplified equation, $C_T = \lambda K$, where $K = A(1 - \gamma A)$, giving Equation 15-1, where $y = C_T$, and $\gamma = -0.0091$.

$$y = \lambda A(1 - \gamma A) \qquad (15\text{-}1)$$

15.4. Step Four, with $y = C_T$

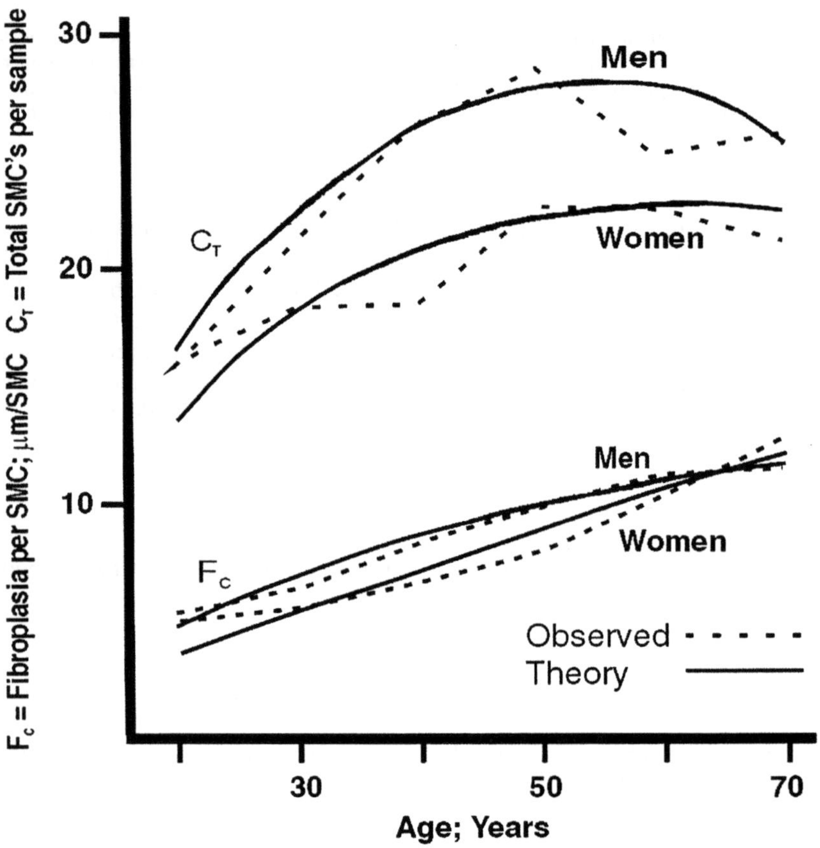

Figure 15-1. Dashed lines present data smoothed by sliding averages of 20-year age groups. Solid curves represent appropriate versions of Equation 15-1 that were fit to the data.

All polynomial terms on age up to the fourth order were rejected for further added inclusion into the equation $C_T = \lambda K$ because they lacked statistical significance in the pools of data for men and women.

15.5. Estimates of λ

With the change of variable K = A(1 - 0.0091A), the λ coefficients of Equation 15-1 were obtained for $y = \lambda K$ by linear regression. These were 1.021 for men and 0.826 for women. Two versions of Equation 15-1 using these values of λ are plotted in Figure 15-1 (upper frame) as solid curves, together with the observed data points as dashed curves. Unlike previous charts, this graph plots data that were smoothed by sliding averages of overlapping 20 year age groups.

15.6. Step One, with $y = F_C$

The intercept term, a, was consistently eliminated.

15.7. Step Two, with $y = F_C$

The fitting of Equation 14.1 to the pool of 357 *basal* men rejected the A^3 term and retained the A and A^2 terms. The fit to the pool of 191 *basal* women rejected the A^2 and the A^3 terms, however, for ease of comparison with the men the insignificant A^2 term ($p = 0.34$) is retained for the fitting of Equation 15-1.

15.8. Step Three, with $y = F_C$

In the expression $F_C = b_1A + b_2A^2$, the ratios b_2/b_1 were close to zero, and statistically significant only for the men. The plots of F_C on age therefore differ little from straight lines (Figure 15-1).

15.9. Step Four, with $y = F_C$

Using $b_2/b_1 = \gamma$, the pools of men and women yielded $\gamma = -0.00580$ and -0.00128 respectively.

15.10. Estimates of λ

With γ taking the foregoing values, and $y = F_C = \lambda K$, λ was estimated by regression, and found to be respectively 0.277 for men and 0.189 for women.

15.11. Taking the product, $F_C * C_T = F$

The foregoing results are summarized in Equations 15-2&3 for men and Equations 15-2a&3a for women.

$$C_T = 1.021A(1 - 0.0091A) \quad (R^2 = 0.869) \tag{15-2}$$

$$C_T = 0.826A(1 - 0.0091A) \quad (R^2 = 0.871) \tag{15-2a}$$

$$F_C = 0.277A(1 - 0.00580A) \quad (R^2 = 0.854) \tag{15-3}$$

$$F_C = 0.189A(1 - 0.00128A) \quad (R^2 = 0.905) \tag{15-3a}$$

The total fibroplastic intimal thickness, F, examined at length in Chapter Fourteen, can be approximated by F ≈ F_C*C_T. This is approximate because F_C sometimes represents the average ratio of F/C_T in 9 coronary sites, and sometimes the ratio of the averages for F and C_T; the average of the ratios is not identical to the ratio of the averages. By taking the product of Equations 15-2 and 15-3, we can write

$$F \approx F_C*C_T = 0.283A^2 - 0.00844A^3 + 0.0000539A^4 = 0.283A^2(1 - 0.0298A) + 0.0000539A^4 \tag{15-4}$$

$$F \approx F_C*C_T = 0.156A^2 - 0.00162A^3 - 0.00000182A^4 = 0.156A^2(1 - 0.0104A) - 0.00000182A^4 \tag{15-4a}$$

15.12. Steps One through Four with y = F

These steps of curve fitting are detailed in Chapter Fourteen. For the pools of men and women in use here, the end results are summarized here.

$$F = 0.230A^2(1 - 0.01057A) \tag{15-5}$$

$$F = 0.126A^2(1 - 0.00825A) \tag{15-5a}$$

The A^4 term is rejected for inclusion in Equations 15-5 and 15-5a as not of statistical significance. These findings confirm Equations 15-5 and 15-5a as satisfactory approximates to Equations 15-4 and 15-4a.

15.13. Two aspects of arterial intimal fibroplasia

Total fibroplastic intimal thickness, F, comprises the two independent aspects, F_C and C_T, the aspect related to thickness per SMC and the aspect imparted by numbers of SMC's. Data on C_T can therefore serve to measure its associated aspect of intimal fibroplasia, a principle that helps to illuminate Equation 4-4 in Chapter Four. In that equation, C_T can be interpreted as substituting for its associated aspect of fibroplasia. Equation 4-5 further emphasizes the principle, because the term F displaces C_T when used conjointly with F_C in predicting atheroma. The two aspects of fibroplasia will receive much attention in the following chapters. These aspects are denoted the F_C-linked component and the C_T-linked component of arterial intimal fibroplasia. Their varied associations with the features of renovasculopathy are informative. It is useful to remember that F_C is the more ominous aspect of fibroplasia. The C_T-linked component of intimal thickening appears to have much weaker associations with atheroma than does the F_C-linked component. (We should also keep in mind that elevated C_T values in the YesA groups are apt to be a consequence of atheroma acting through biased censoring and not necessarily a cause of atheroma. It may be a marker for arteries that are predisposed to ready conversion of fibroplasia to atheroma.)

An important detail in the patterns of Figure 15-1 is that C_T has little variation with aging after 30 years of age. On the other hand, F_C rises steadily into old age. Hence, the variable that dominates atheroma formation is also the variable that dominates the changes that accompany aging. The impact of SMC numbers on intimal thickness is largely a matter of growth and maturation, with only small components added after age 30 years and later subtracted after age 60 years.

15.14. Estimating normal F_C

Equations 15-3&3a yield F_C at age 20 years of 2.7 μm/SMC for women and 3.9 μm/SMC for men, with an average of 3.3 μm/SMC. This average value is used in Chapter One for estimating the normal values of F_C in adults at the onset of the fibroplastic aging process.

CHAPTER SIXTEEN

FIBROPLASIA PER SMC IN THE MEDIA OF CORONARY ARTERIES

Abstract. Foregoing chapters repeatedly emphasized the importance of F_C, fibroplasia per SMC in the intimal layer, in the formation of atheroma. This chapter now reviews evidence that the same principles apply also in the medial layer. The evidence suggests a generalized arterial degeneration acting throughout the artery wall, but progressing much faster in the intima.

The data reported here provide support for a simple hypothesis: The principal aging effect leading to atheroma is taken to be the expansion of fibroplastic realms that the SMC's build for themselves to occupy. The expansion is viewed as a generalized process affecting the medial and intimal layers simultaneously. The fibroplasia per SMC in the media begins to expand around the same age as in the intima, but does not progress as rapidly as it does in the intima. Some parts of the intima quickly surpass the threshold of 9.3 µm/SMC to form atheromas, while all parts of the media fail to attain this magnitude, even in extreme old age. Hence only the intima manifests aggressive fibroplastic arteriosclerosis, while the media remains affected only by the quiescent form of fibroplasia. This hypothesis offers a possible explanation for avoidance of the medial layer by atheromas.

16.1. Division of the cases at the median of medial thickness

The cases under study here are the same as in Chapter One, but with increased numbers added by subsequent ongoing accumulation of more specimens. In all *basal* cases,

Table 16-1. Construction of groups above (High) and below (Low) the 10-year age group medians of medial thickness; *basal* cases.†

Age; Years	Median of Medial Thickness µm	N; Number of cases: Women Low	Women High	Men Low	Men High	Percent above median: Women	Men	Mean thickness of Media; µm: Women	Men
15-24	147	13	8	19	23	38	55	145	158
25-34	175	19	8	19	31	30	62	168	174
35-44	170	23	7	26	41	23	61	161	178
45-54	193	16	6	20	31	27	61	192	196
55-64	190	5	7	10	9	58	47	185	206
65-79	206	5	10	11	6	67	35	211	191

† Modified from reference [56] with permission from Elsevier Science.

medial thickness was measured at nine sites in NoA specimens, and at fewer than nine sites in YesA cases because of excluding sites with atheromas. SMC numbers were counted in a 100 μm wide band through the medial thickness. Mean thickness is M, mean number of SMC's is C_M, and their ratio is F_M, fibroplastic thickness per SMC. Within each 10 year age group, the subjects were ranked on medial thickness and the median of each group was identified; High M cases are those falling above the median and Low M cases fall below the median. Table 16-1 gives the cutoff points for each age group. It is presumed here, as in Article 14.12, that the high groups and low groups can be treated as cohorts that share these high or low traits throughout life from one age group to another.

At ages 15-54 years, most of the women were in the Low M groups and most of the men in the High M groups. These patterns reversed at ages 55-79 years. In women, mean thickness of media increased slowly from 145 μm at ages 15-24 years

Table 16-2. Presence of atheroma in men and women of Low and High classification by medial thickness; *basal* cases.*†

Age; Years	Y; YesA: percent with Atheroma			
	Women		Men	
	Low M	High M	Low M	High M
15-24	8	0	5	0
25-34	0	13	0	19
35-44	9	0	42	46
45-54	0	67	50	68
55-64	60	71	60	100
65-79	80	80	82	83
Unweighted mean	25	38	40	53
	A	AB	B	C

ANOVA	F	p
Sex	9.1	**0.00**
Age	30.1	**0.00**
Media group	7.3	**0.00**
Sex*Age	2.6	**0.02**
Sex*Media	0.0	0.95
Age*Media	3.1	**0.01**
Sex*Age*Media	1.4	0.21

* Marginal means failing to share a symbol A, B or C differ significantly, $p < 0.05$.
† Modified from reference [55] with permission from Elsevier Science Science

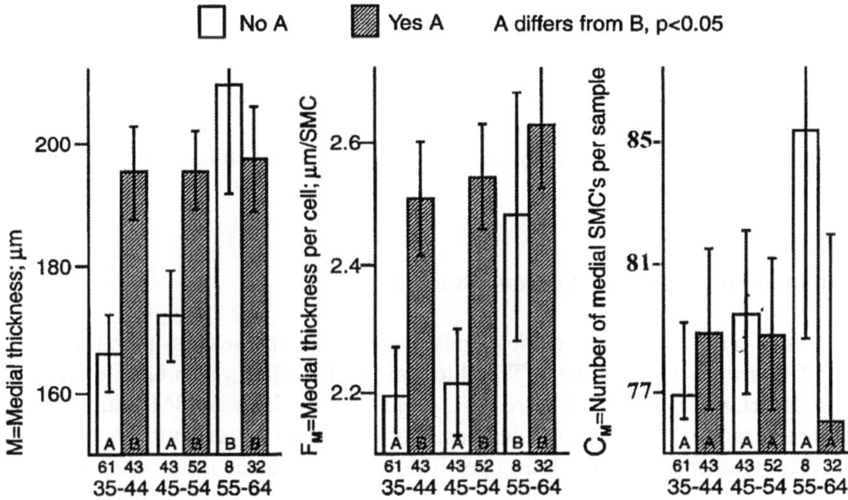

Figure 16-1. Means and standard errors of the means for selected characteristics of coronary medial layer according to presence (black bars) or absence (white bars) of atheroma in 10-year age groups. Basal men and women. For each of the three variables, columns that fail to share a symbol, A or B, differ significantly at $p < 0.05$. Numbers of cases at bases of columns. Reproduced from reference [56] with permission from Elsevier Science

to 161 µm at ages 35-44 years, then rose swiftly toward 211 µm at ages 65-79 years. In men, mean medial thickness of 158 µm at ages 15-24 years exceeded that of women, but subsequently rose only modestly to a nearly level plateau of 178-206 µm at ages 35-79 years.

16.2. Atheroma in Low M and High M groups

A greater proportion of subjects were scored as having atheroma (YesA) in the High M than in the Low M groups (Table 16-2), by 38%:25% in women and 53%:40% in men overall ($p < 0.05$ binomial probability, also significant by ANOVA).

16.3. Components of medial thickness

As earlier noted in the intima, for the media also thickness comprises SMC numbers, C_M, and their average associated matrix, F_M. Equation 16-1 shows these two components to contribute about equally to variations among cases in medial thickness. This result can be seen here, because the two variables have standardized coefficients in the discriminant function that are about equal to each other.

$$W = 3.12F_M + 3.20C_M \quad (T^2 = 5.42 \quad \% = 87 \quad c=-25.7) \qquad (16\text{-}1)$$

The coefficients on the right side of the equation are determined by discriminant function analysis (symbols are explained in Article 1.7, Chapter One); High M subjects are optimally distinguished from Low M subjects when W = 0. The equality of weights for these two components is wholly lost, however, when the emergence of atheroma is explored, as seen next.

16.4. Relationship of medial components to atheroma

The media of YesA coronary arteries tends toward greater thickness than that of NoA arteries (Figure 16-1, left panel). The thickening of media is attributable entirely to increased thickness per SMC (Figure 16-1, middle panel), because SMC numbers show no tendency to vary significantly between subjects with or without atheroma (right panel). The greater medial thickness per SMC in YesA cases is conspicuous only at ages 35-54 years, because subjects less than age 35 years seldom have atheroma, and those over age 54 years seldom lack atheroma. Inclusion of sex in the statistical model does not alter these conclusions (data not shown), indicating that the relationships are alike for the averages in men and women.

The variable F_M averages in the range 2.16 to 2.63 µm/SMC (middle panel). (This low range is seldom encountered in the intima, even of youthful subjects.) The highest average value obtained in any of these 370 specimens was 4.2 µm/SMC for medial thickness per SMC. Hence no subject had the mean F_M approach at all close to the threshold for atheroma that obtains in the intima, 9.3 µm/SMC.

16.5. Intimal fibroplasia in Low M and High M groups

Intimal fibroplastic thickness is greater in the High M groups than in the Low M groups, by an overall average of about 26 % (Table 16-3). Low M groups display aging patterns similar to those of High M groups. Intimal fibroplasia in women rises slowly from ages 15 to 44 years, and more swiftly thereafter, maintaining at all ages a greater intimal thickness in the High M groups of about 60 µm on average. Intimal fibroplasia in men rises rapidly from ages 15 to 54 years, and very slowly thereafter, maintaining throughout all but the oldest age groups an intimal thickness averaging about 36 µm greater in the High M groups.

16.6. Components of fibroplastic intimal thickening

Table 16-3. Mean fibroplastic intimal thickness in men and women of Low and High classification by medial thickness; *basal* cases.*†

Age; Years	F_I; Fibroplastic intima total; µm			
	Women		Men	
	Low M	High M	Low M	High M
15-24	65	119	68	97
25-34	87	137	95	143
35-44	115	130	204	230
45-54	161	211	232	282
55-64	141	293	254	283
65-79	254	282	293	282
Unweighted mean	135	194	191	219
	A	BC	B	C

ANOVA	F	p
Sex	10.2	**0.00**
Age	24.2	**0.00**
Media group	12.5	**0.00**
Sex*Age	2.3	**0.04**
Sex*Media	1.4	0.23
Age*Media	0.7	0.63
Sex*Age*Media	0.6	0.73

* Marginal means failing to share a symbol A, B or C differ significantly, $p < 0.05$.
† Modified from reference [56] with permission from Elsevier Science.

Intimal SMC numbers, C_T, in High M groups exceed those in Low M groups at ages 15-44 years, but not at ages 45-79 years (Table 16-4). C_T in men tends to exceed that in women at some ages, but only because more men occupy the High M groups; the sex difference loses statistical significance after matching for medial thickness in Table 16-4. Intimal fibroplasia per SMC, F_C, is equal or greater in High M groups in every comparison with Low M groups of men and women at all ages, by an overall average of about 1.7 µm/SMC. Men continue to exceed women in F_C after matching for medial thickness, the sex difference retaining statistical significance after matching in Table 16-4.

16.7. Photographic examples

Table 16-4. Numbers of intimal SMC's and fibroplastic thickness per SMC in men and women of Low and High classification by medial thickness; *basal* cases.*†

Age; years	F_C; Fibroplasia per SMC; µm/SMC				C_T; Numbers of SMC's; SMC's/sample			
	Women		Men		Women		Men	
	Low M	High M	Low M	High M	Low M	High M	Low M	High M
15-24	4.8	5.6	4.6	5.7	13.5	21.5	15.2	17.6
25-34	5.2	5.9	4.9	6.6	17.2	22.6	19.3	22.3
35-44	6.5	6.1	8.2	8.4	17.6	21.3	24.8	27.3
45-54	6.2	10.0	9.2	9.7	27.1	21.7	25.5	31.2
55-64	7.1	10.7	8.8	15.5	21.6	26.7	28.2	20.4
65-79	11.4	13.2	11.1	13.2	24.8	21.5	27.2	22.0
Unweighted mean	6.5 A	8.5 B	7.7 B	9.8 C	20.7 A	22.6 A	23.7 A	23.5 A

ANOVA	F	p	All ages F	All ages p	Ages 15-44 years F	Ages 15-44 years p
Sex	7.9	**0.01**	3.5	0.06	2.4	0.12
Age	33.2	**0.00**	6.6	**0.00**	5.6	**0.00**
Media group	27.5	**0.00**	0.8	0.36	8.8	**0.00**
Sex*Age	2.1	**0.06**	1.3	0.27	2.8	**0.07**
Sex*Media	0.3	0.57	1.1	0.29	1.2	0.28
Age*Media	3.5	**0.00**	1.7	0.14	0.2	0.83
Sex*Age*Media	1.4	0.24	1.9	0.10	0.2	0.81

* Unweighted marginal means that fail to share a symbol, A, B, or C differ significantly, $p < 0.05$.
† Modified from reference [56] with permission from Elsevier Science.

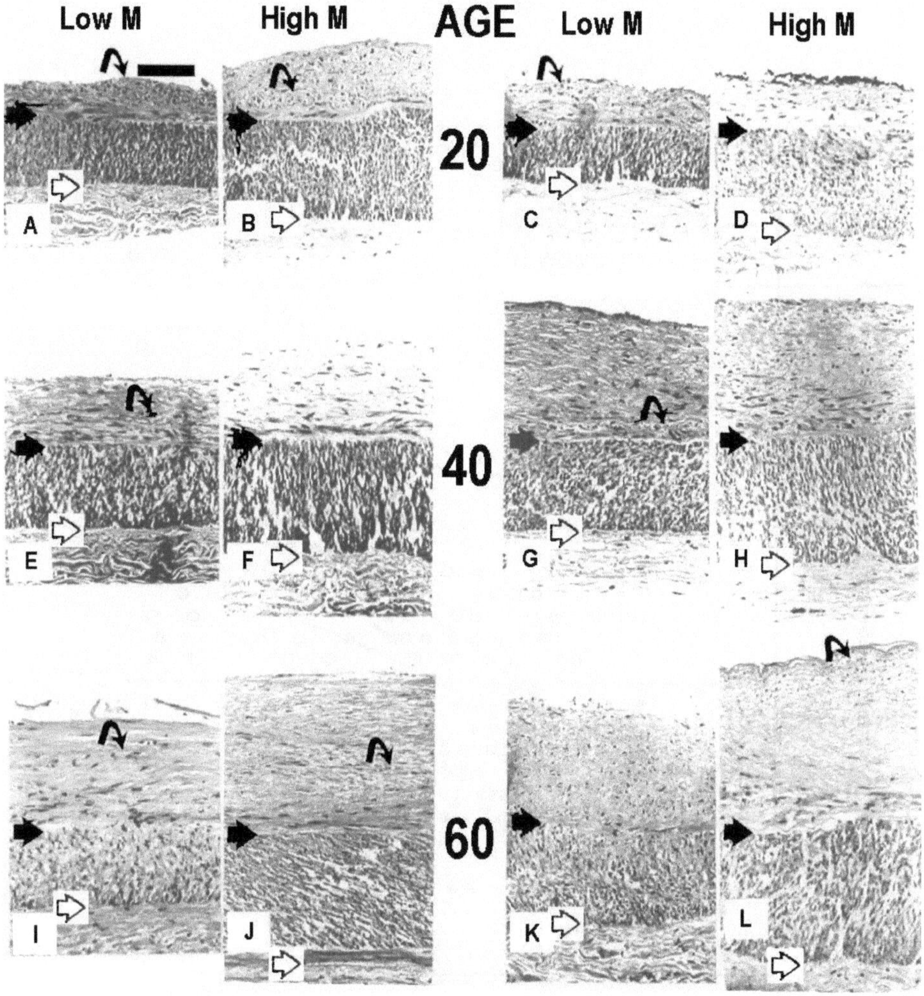

Figure 16-2. "Typical" examples chosen to show here have medial and intimal quantities that fall near the averages in Tables 16-1,2,3,&4 and in Figure 16-1. The six panels on the left represent women, those on the right represent men. Black arrows mark intima-media boundaries, clear arrows mark media-adventitia boundaries. Bent arrows mark islands of media-like intima. H&E, bar = 100 μm applies to all frames. Reproduced from reference [56] with permission from Elsevier Science.

Photographs in Figure 16-2 exemplify typical appearances in Low M and High M men and women of increasing ages. These are "typical" in the sense of displaying medial and intimal thicknesses near the averages in Tables 16-1,2,3,&4 and in Figure 16-1. In Figure 16-2, the photos of High M examples seem almost like viewing the Low M

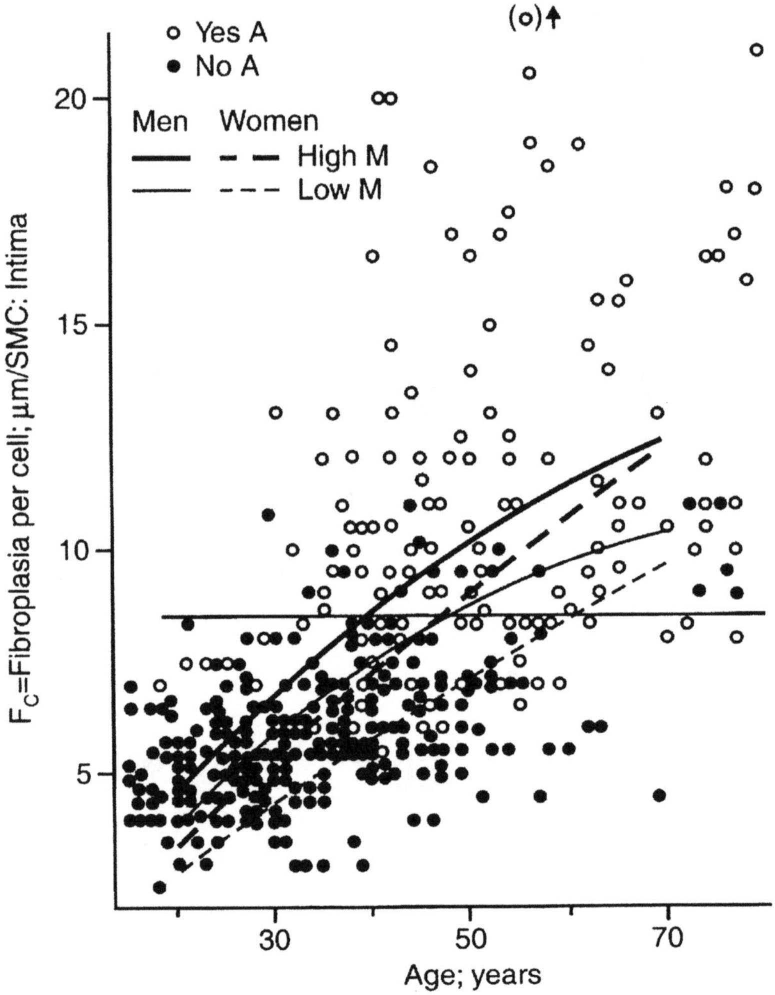

Figure 16-3. Each symbol represents an autopsy with presence (circles) or absence (dots) of atheroma in the coronary specimen. Curves represent regression lines fit to men (solid) and women (dashed) in the High M (broad) and Low M (narrow) groups. Horizontal line at 8.6 µm/SMC represents the discriminant function that optimally separates circles from dots. Arrow directed upward identifies an off scale outlier case with F_C = 32 µm/SMC. Basal men and women. Reproduced from reference [56] with permission from Elsevier Science.

examples through a 1.5X magnifying glass. The difference, however, holds greater importance than merely scaling the units of measure, because the High M categories are at greater risk of acquiring an atheroma (Table 16-2). Note the tendency for SMC's to spread apart from each other in the High M compared with the Low M examples, as we previously saw them do in aging intima.

16.8. Intimal fibroplasia and the intrusion of atheroma

We have previously reviewed evidence that fibroplasia per SMC in the coronary intima is the principal aging effect leading to atheroma. Figure 16-3 illustrates this principle graphically. As the artery ages, F_C shows an average tendency to increase. The subjects who increase beyond a certain threshold abruptly manifest a high probability of acquiring an atheroma. An equation derived by discriminant function analysis determines the threshold to be 8.6 μm/SMC in this body of data, and this is plotted as a horizontal line in Figure 16-3. This compares with values ranging near 9.3 μm/SMC, encountered repeatedly in Chapter One and subsequent chapters. The solitary horizontal line depicts a discriminant function, and that function is plotted as a single horizontal line because it rejects age, sex, and medial thickness as not significant. These results indicate that the same threshold applies alike to men and women of all ages and medial thicknesses in this data set. Men and women who are above average in medial thickness tend to reach the threshold sooner than the below average subjects. Men tend to reach the threshold sooner than women, irrespective of medial thickness. The pattern implies that we know what aging does to prepare the coronary intima for atheroma, and that we have found maleness and medial thickness to act solely by modifying the simple aging effect.

These results are shown by the regression lines plotted as curves in Figure 16-3. The solid curves representing High M and Low M men cross the threshold near ages 39 and 47 years respectively, about 20 % later for the Low M groups. The dashed curves representing High M and Low M women cross near ages 46 and 60 years respectively, about 30 % later for the Low M groups. This early arrival of F_C at the threshold fully "explains" statistically the greater tendency to atheroma in the High M groups without invoking any other cause. The same principle applies to the sex difference; F_C progresses faster in men so that they arrive sooner at the threshold, and this early arrival can fully "explain" the greater propensity of men to atheroma in the coronary artery. The magnitude of the medial grouping effect is almost as great as the sex difference.

16.9. The course of F_C on age in Low M and High M groups

Graphs of regression lines like those in Figure 16-3 are plotted in Figure15-1 of Chapter Fifteen, without separation of subjects by medial thickness. Those earlier graphs represent Equation 15-3, of the form $y = \lambda A(1 - \gamma A)$. Following the development of Chapter Fifteen with $y = F_C$, these regression equations were determined here for

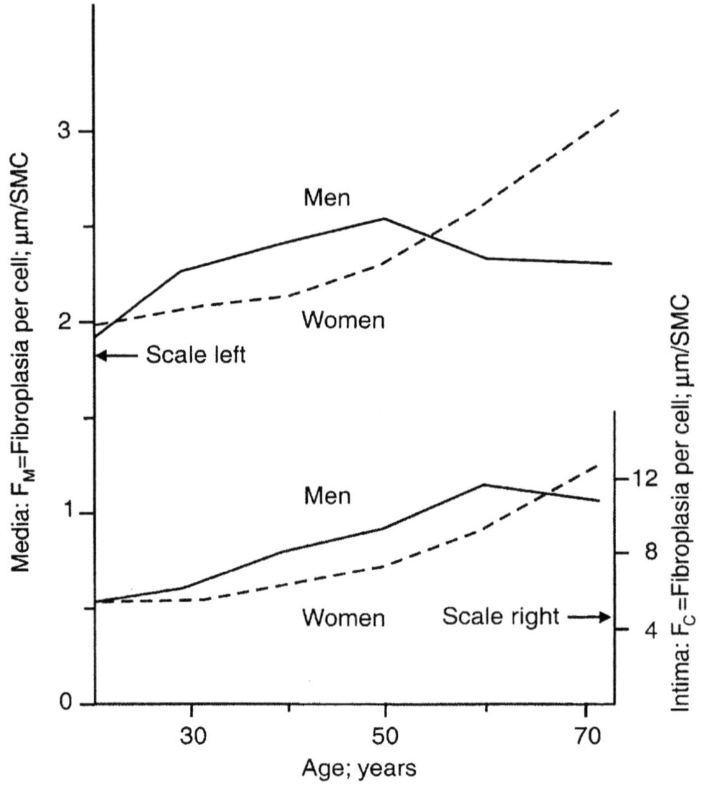

Figure 16-4. Fibroplastic thickness per SMC in the media (upper curves, scale left) and intima (lower curves, scale right); means within 10-year age groups joined by solid lines (men) and dashed lines (women). Basal men and women. Reproduced from reference [56] with permission from Elsevier Science.

men and women in the Low M and High M groups. The parameter γ fell near the previously observed value of -0.00580 for men of both media classes. Hence substituting the average of the new estimates, γ = -0.00506, allowed estimation of λ which became 0.243 for the Low M men and 0.282 for the High M men. For women of both media classes, estimates of γ were similar and averaged γ = -0.000708, not greatly different from γ = -0.00128 obtained in Article 15.9. Using the new estimates allows λ to be found; it is 0.166 for the Low M women and 0.209 for the High M women. The small value of γ = -0.000708 introduces a barely perceptible curvature in the plots of Figure 16-3, in keeping with its lack of statistical significance.

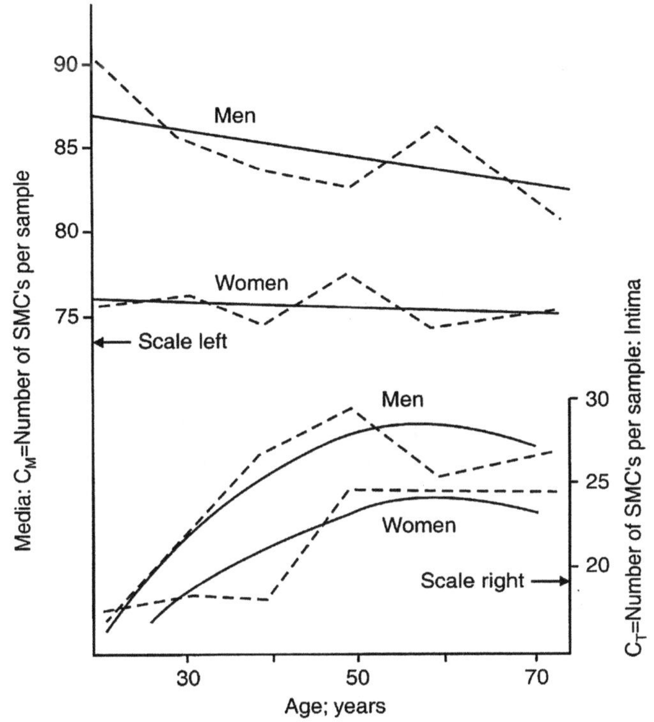

Figure 16-5. SMC numbers in media (upper lines, scale left) and intima (lower curves, scale right); means within 10-year age groups are joined by dashed lines. Solid lines and curves represent regression lines. Reproduced from reference [56] with permission from Elsevier Science.

16.10. Fibroplasia per SMC in media and intima

Aging tends to expand the bulk of tissue associated with each SMC in patterns that differ between men and women (Figure 16-4). The patterns for the media, F_M, show a general resemblance to those for the intima, F_C. Both F_C and F_M increase rapidly in men from ages 15 to 54 years, while the women lag. From ages 55 to 79 years, both F_C and F_M accelerate in women but lag in men.

16.11. Numbers of SMC's in media and intima

Numbers of SMC's in the media tend to decline with age, faster in men than in women (Figure 16-5, upper lines). Numbers of SMC's in the intima tend to increase from ages

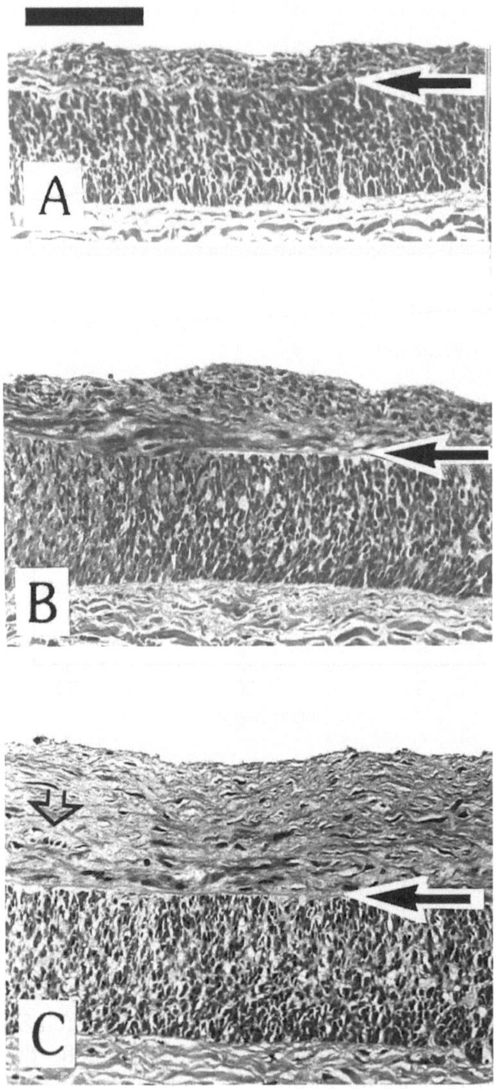

Figure 16-6. Views of coronary artery from a 21 year old woman who died of violence. A: Early intima is often media-like, with closely spaced circumferentially oriented SMC's. B: Longitudinal reorientation of SMC's in maturing intima is often asynchronous from place to place. C. Mature intima with longitudinal SMC's more widely separated than in the media often retain small clusters of SMC's forming media-like islands (hollow arrow). Solid arrows mark intima-media boundaries. H&E, bar = 100 μm. Reproduced from reference [56] with permission from Elsevier Science.

15 to 54 years, and then to stabilize or decline slightly to age 79 years (Figure 16-5, lower lines); again, the increase is greater in men than in women. The augmenting of SMC numbers in the intima exceeds the depletion in the media up to age 50 years so that total numbers increase slightly. After age 54 years, indications are for a slow decline in total SMC's in the entire artery wall.

Regression lines fit to the data and plotted in Figure 16-5 are obtained by following the development of Chapter Fifteen. For men and women respectively, $C_M = 90.6 - 0.183A$ and $C_M = 77.2 - 0.039A$; $C_T = 1.003A - 0.00880A^2$ and $C_T = 0.836A - 0.00723A^2$, with the intercept rejected in all settings.

16.12. The interaction of medial and intimal SMC's seen visually

The normal growth of coronary artery is thought to be associated with the formation and later refilling of gaps in the internal elastica [23], which allows the artery to acquire a larger caliber. Through these gaps, medial cells can relocate to the inner side of the elastica to give rise to the intima [43]. A comparable sequence of events seems to occur in pathological enlargements of arteries, as in experimental arteriovenous fistula [30], and other systems [38]. Something similar might be happening as arteries become progressively dilated, tortuous, and rigid in the aging process [15,25,42], a process which might be called "presbyectasia" for want of a name.

The newly forming intima of young specimens often has a media-like structure, with closely spaced circumferential SMC's (Figure 16-6A). This structure quickly alters by rotation of SMC's into a longitudinal orientation with construction of bulky matrix masses to form the mature intima (Figures 16-6B&C). Media-like structures can sometimes persist as isolated islands, even late in life (bent arrows, Figure 16-2). It seems plausible to suggest that these media-like islands may contribute a degree of resistance to atheroma. We have repeatedly encountered evidence in past chapters that atheromas favor sites with few SMC's, which is to say that they avoid sites with numerous SMC's, especially those in clusters of unusually many cells. Residual media-like islands would have both of these protective properties.

16.13. The assessment of intimal and medial thicknesses among human populations

Chapter Fourteen reported findings in coronary arteries from eight population groups (cf. Article 14.13). SMC numbers were inaccessible for technical reasons, and only thickness measurements can be evaluated. Figures 16-7&8 summarize some especially cogent findings on medial and intimal thicknesses of coronary arteries.

The subjects of that study were divided into "young" and "old" at age 30 years. The issue of interest was the "tracking" of variables from younger into older ages. Symbols in Figures 16-7&8 contain numerals which serve to identify the

Table 16-5. Spearman rank order correlation coefficients relating selected variables at ages 15-29 years to those at 30-69 years in 16 population groupings.†

Variable at ages 15-29	Variables at ages 30-69 years				
	F_o	M_o	T_o	Y_o	P_o
F_y	0.56*	0.64**	0.69**	0.42	0.34
M_y	0.61*	0.79**	0.72**	0.51*	0.49
T_y	0.56*	0.73**	0.69**	0.43	0.39
Y_y	0.04	0.26	0.17	0.41	0.10
P_y	0.20	0.23	0.24	0.40	0.33

† F is fibroplastic intimal thickness; M is medial thickness; T is total wall, F+M; Y is % of cases classes as YesA; P is Percent of nonnecrotic specimen having fatty streak elements; subscript y is young and o is old.
*p<0.05,**p<0.01.
† Reproduced from reference [55] with permission from Elsevier Science.

populations by their "reference ranks". This ranking refers to the extent of grossly assessed raised lesions found by visual inspection of coronary arteries and aortas from those populations. The populations in order of the rankings are 1 - New Orleans (White), 2 - Durban (Indian), 3 - New Orleans (Black), 4 - Manila, 5 - Sao Paulo (White), 6 - Mexico City, 7 - Bogota, and 8 - Durban (Bantu). Treating men and women separately generates a total of 16 population groups.

These rankings indicate that the New Orleans (White) group had the most extensive involvement of arteries by visible raised lesions, while Durban (Bantu) had the least extensive involvement in *basal* subjects (unweighted averages of 10 year age groups from 10 to 69 years, men and women combined).

Fibroplastic intimal thickness in the young people of a population, F_y, significantly foreshadows fibroplasia in the older subjects, F_o. Figure 16-7 illustrates this conclusion graphically and Table 16-5 shows the significance in the form of a rank correlation of r = 0.56 (p<0.05). Medial thickness also shows the same trend, plotted in Figure 16-8 and with rank correlation of 0.79 (p<0.01). Youthful intima foreshadows elderly media (r=0.64, p<0.01) and youthful media foreshadows elderly intima (r=0.61, p<0.05). Only the medial thickness foreshadows elderly atheroma to a significant magnitude (r=0.51, p<0.05).

The findings in Table 16-5 serve to reinforce the conclusions previously noted.

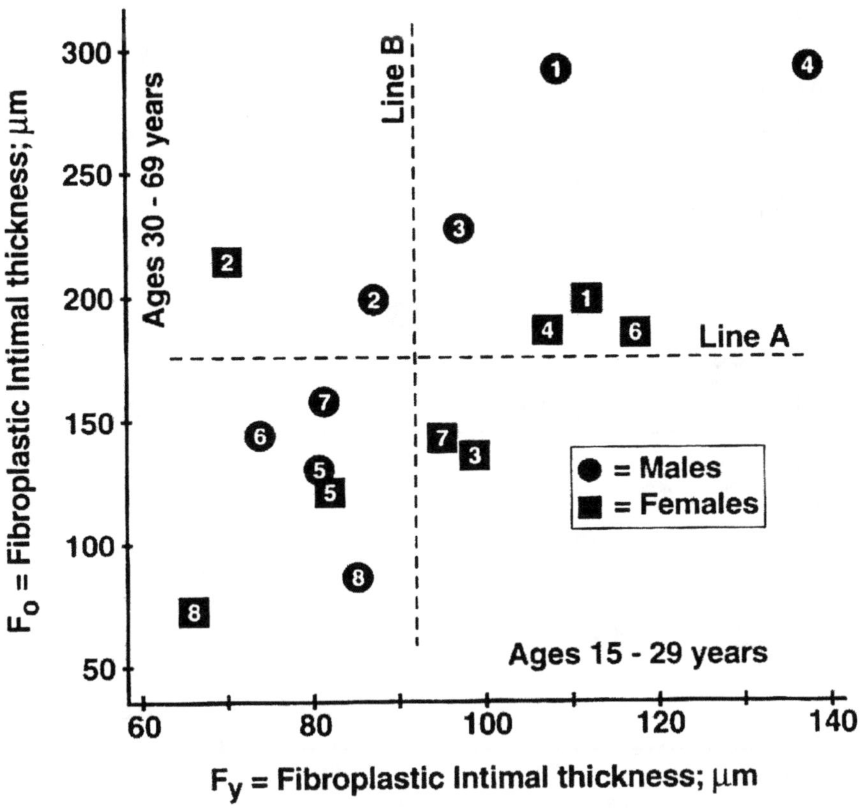

Figure 16-7. Means of the variable F, age-adjusted within each age range, are plotted against each other. Numbers within the symbols identify the 8 geographic locations that are represented (cf. Article 16.13). The median of the 16 values on the horizontal is marked by line B, and on the vertical by line A. Reproduced from reference [55] with permission from Elsevier Science.

The thicknesses of intima and media tend to be locked together in a way that suggests a generalized process affecting all parts of the arterial wall simultaneously. They furthermore add significance to the association of medial thickness in youth with late life atheroma.

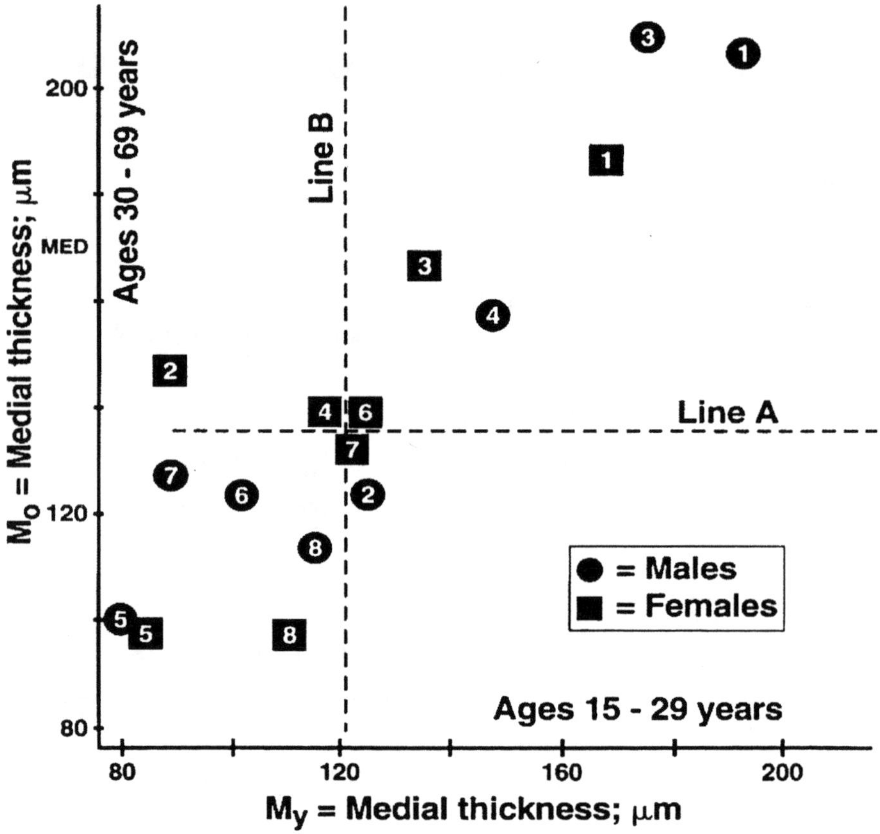

Figure 16-8. As in Figure 16-7 for the variable M.. Reproduced from reference [55] with permission from Elsevier Science.

Table 16A. Descriptive statistics for the variables in this chapter, including product moment correlation coefficients; 370 *basal* men and women of New Orleans aged 15-79 years. Reproduced from reference [56] with permission from Elsevier Science.†

Variable					Correlation coefficients						
Name	Symbol	Units	Mean	SD	M	F_{M}	C_{M}	F	F_{C}	C_{T}	Y
Age	A	Years	39.8	15.0	.31**	.40**	-.12*	.54**	.57**	.29**	.51**
Media	M	μm	178	49.3	.---	.64**	.45**	.40**	.40**	.26**	.36**
Fibroplasia Per SMC	F_M	μm/SMC	2.26	0.61	.61**	.---	-.36**	.38**	.44**	.18**	.36**
SMC's	C_M	$N/(100\mu m)^2$	80.8	18.1	.51**	-.35**	.---	.04	-.03	.13*	-.02
Intima	F	μm	179	122	.31**	.22**	.13*	.---	.72**	.68**	.66**
Fibroplasia Per SMC	F_C	μm/SMC	7.73	3.78	.30**	.28**	.05	.61**	.---	.07	.68**
SMC's	C_T	$N/(100\mu m)^2$	22.9	10.2	.22**	.06	.18**	.66**	-.08	.---	.24**
YesA; No=0 Yes=100	Y	100,0	29.1	45.5	.24**	.20**	.04	.51**	.54**	.13*	.---

† Raw correlations in upper right triangle, age adjusted partial correlations in lower left triangle.
* = Differs from zero at $p < 0.05$, ** at $p<0.01$.

CHAPTER SEVENTEEN

INFLUENCE OF ARTERIOLAR HYALINIZATION ON RENOVASCULAR FIBROPLASIA

Abstract. Hyaline deposits in arterioles serve as indicators for a state of accelerated arterial intimal fibroplasia in the renal cortex. The acceleration seems to be a speeding of the aging process, because of how it enters into the formula that describes the way aging influences arterial intima. The "aging risk factor" therefore emerges as a quantity that can vary among individuals in a way that confers visible indicators in autopsy tissues.

Hyaline deposits are often seen in the microvasculature of renal cortex. These are readily revealed in paraffin sections with the PAS stain (Figure 17-1). They are most numerous in afferent arterioles, but also often attack efferent arterioles and terminal arteries (<80 μm outer diameter). These deposits are quantified by counting the numbers of affected vessels per cm^2 of observed area of renal cortex (call this Hy when expressed as the square root to diminish the severe skewness of the distribution). Specimens with numerous hyalinized vessels, Hy, have an average tendency to express severe arterial intimal fibroplasia at the conduit, R_c, and resistance, R_r, levels. To simplify the following discussion, the variable R_a is introduced, $R_a = (R_c + R_r)/2$. The autopsy cases examined in this inquiry are described in Chapter One, Article 1.9, and are the same as used in Chapter Fourteen.

17.1. Constructing tertiles of Hy

Within each 10-year age group, *basal* subjects were ranked on Hy, and divided at the 33⅓ and 66⅔ percentiles to construct Low Hy, Mid Hy, and High Hy subgroups with numbers as nearly equal as possible. In ages 15-24 years, 38 subjects had Hy = 0 and these filled the Low Hy group; a division of the remainder into 13 Mid and 10 High subjects was as near to equal as possible. The cutoff points for all 10-year age groups were obtained in this way and are given in Table 17-1 along with the means of Hy. It is assumed again, as in Article 14.12, that the High Hy groups represent cohorts that share the property of severe arteriolar hyalinization as an ongoing lifelong characteristic, at least as a trend for group averages. Table 17-1 shows the cutoff points and the means of Hy to vary between age groups, rising in a sigmoid pattern to a peak at ages 55-64 years, and falling thereafter. The approximate pattern is to increase rapidly in early life and then to maintain a nearly stable level after age 34 years in these data.

17.2. Renovascular fibroplasia, R_a, in tertiles of Hy

The rates of increase of R_a with age are low, intermediate, or high in keeping with the Hy groupings (Table 17-1). The trend is statistically significant within 10-year age

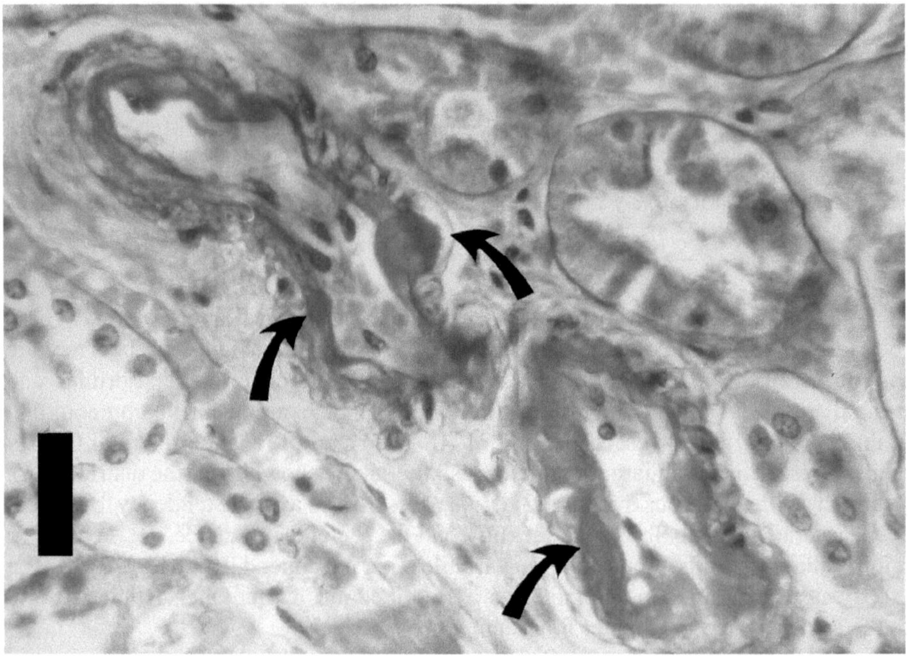

Figure 17-1. Hyaline deposits in an afferent arteriole; these are quantitated by counting affected vessels in a unit area of tissue section. Multiple deposits (arrows) in a single arteriole yields a count of "one". Bar = 25 μm, PAS-Alcian blue paraffin section.

ranges after age 34 years, and is consistent as early as 15-24 years. The High Hy groups in each of the 10-year age ranges all display accelerated progression of arterial fibroplasia compared with Low Hy and Mid Hy groups. The magnitude of the acceleration relative to the Low Hy groups is 35 % in the age range 15-24 years, and is respectively 66, 31, 41, 27, and 20 % in ages 25-34, 35-44, 45-54, 55-64, and 65-79 years. The average of 37 % emphasizes that the effect is weak. Fibroplasia progresses in the subjects with minimal hyalinization at about 2.1 %od per decade, and accelerates only modestly with severe hyalinization to about 2.5 %od per decade.

17.3. Constructing tertiles of R_a

As for Hy, groups were constructed for Low R_a, Mid R_a, and High R_a within 10-year age groups. The means and cutoff points for R_a are given in Table 17-2. R_a rises with age in a pattern different from that for Hy. The increase fails to plateau in mid life, but instead continues into old age, following sigmoid curves like those in Chapter Fourteen.

Table 17-1. Cutoff points of Hy by 10-year age group, and means of Hy and R_a within tertiles of Low, Mid, and High renal arteriolar hyalinization, Hy; pool of all *basal* subjects.*

Age; years	Hy cutoff point Low to Mid	Hy cutoff point Mid to High	Hy; $\sqrt{(N/cm^2)}$ Low	Hy Mid	Hy High	R_a; %od Low	R_a Mid	R_a High	Number of cases Low	Number of cases Mid	Number of cases High
15-24	0.0	0.9	0.0	0.8	1.6	2.0 A	2.2 A	2.7 A	38	13	10
25-34	0.8	1.8	0.2	1.3	3.1	3.0 A	3.4 A	5.0 A	25	28	24
35-44	1.3	2.8	0.6	2.0	4.9	5.9 A	5.9 A	7.7 B	30	33	33
45-54	1.3	2.8	0.5	1.9	5.0	7.4 A	9.3 B	10.4 B	24	23	26
55-64	1.8	3.8	0.9	2.6	5.3	10.2 A	12.6 A.B	13.0 B	10	10	11
65-79	1.4	2.8	0.5	2.0	4.8	12.7 A	14.0 A	15.3 A	11	9	10
Unweighted Mean						6.9 A	7.9 B	9.0 C			

ANOVA	F	.(p)
Age	80.6	(0.00)
Hyaline	10.9	(0.00)
Interactions	0.6	(0.83)

* Entries within a row that fail to share a symbol A or B differ significantly at $p < 0.05$.

17.4. Progression of Hy by tertiles of R_a

Hy in the High R_a groups exceeds that in the Low R_a groups at all ages (Table 17-2). The difference is already established nearly in full by ages 25-34 years, with little or no further divergence of Hy between R_a groups in later life. The two kinds of renovasculopathy are not fully symmetrical in their effects upon each other. The graphs

Table 17-2. Cutoff points for R_a by age, and means of R_a and Hy by tertiles of renal arterial fibroplasia, R_a, in 10-year age groups; pool of all *basal* subjects.*

Age;	R_a cutoff point		R_a %od			Hy;√(N/cm²)			Number of cases		
years	LowtoMid	MidtoHigh	Low	Mid	High	Low	Mid	High	Low	Mid	High
15-24	1.6	2.8	0.7	2.1	3.6	0.2 A	0.4 A	0.6 A	20	19	22
25-34	2.1	4.5	1.4	3.2	6.7	1.0 A	1.4 A.B	2.2 .B	26	26	25
35-44	4.4	7.7	3.1	5.9	10.3	1.7 A	2.5 A.B	3.4 .B	32	31	33
45-54	6.6	10.6	4.5	9.1	13.9	2.1 A	2.4 A.B	3.2 .B	25	24	24
55-64	9.8	13.5	8.5	11.7	15.3	2.5 A	2.7 A	3.8 A	10	10	11
65-79	12.5	17.0	8.2	15.2	19.5	1.6 A	2.8 A	2.9 A	11	10	9
Unweighted Mean						1.5 A	2.0 B	2.7 C			

ANOVA	F	(p)
Age	13.6	(0.00)
Hyaline	9.2	(0.00)
Interactions	0.4	(0.95)

* Entries within a row that fail to share a symbol A or B differ significantly at $p < 0.05$.

in Figure 17-2 serve to illustrate these observations by plotting data from Tables 17-1&2, along with theoretical fits to the curves.

17.5. Curve fitting, R_x on age

The development here follows that of Chapters Fourteen and Fifteen. Step One of the curve fitting begins as before with Equation 14-1, $y = b_1A + b_2A^2 + b_3A^3 + a$, where y refers to a measure of fibroplasia in an arterial segment of a group of *basal* subjects

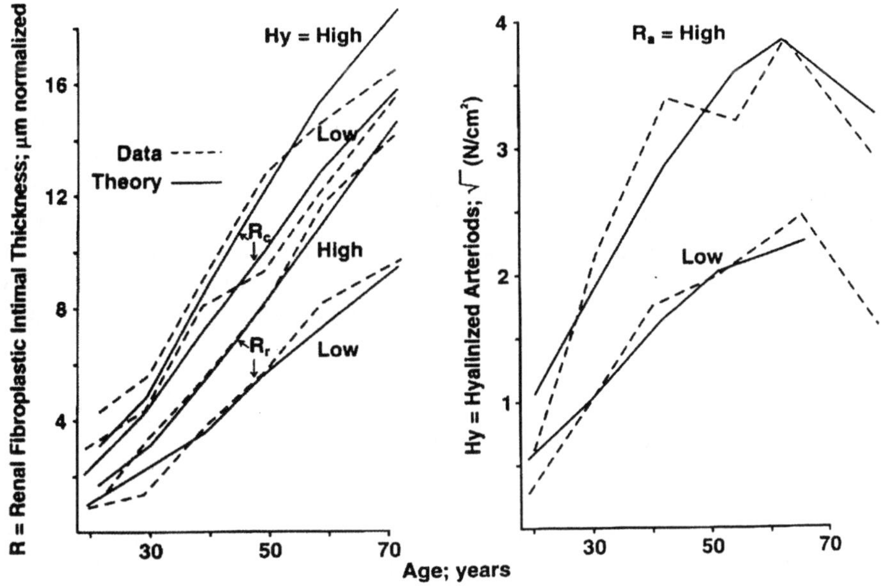

Figure 17-2. Means of Hy in the High R_a and Low R_a subgroups (right panel) are taken from Table 17-2. Data in the left panel, taken from Table 17-1, are restructured to separate R_a into its components R_r and R_c for plotting within the High Hy and Low Hy subgroups. Dashed curves represent the observed data and the solid curves represent the versions of Equation 14-2 fit by regression to the respective data sets. For this plot, the values of R_c are multiplied by 1.5 to convert units from %od into µm of intimal thickness normalized to an artery of 150 µm outer diameter, and R_r by 1.0 to normalize to an artery of 100 µm od.

aged 15-79 years and A is age in years. The fitting was done in the three groupings of Low Hy, Mid Hy, and High Hy for the two arterial segments, renal resistance arteries ($y = R_r$) and renal conduit arteries ($y = R_c$). In all six fittings, the intercept term, a, and the linear term on age, b_1A, were insignificant or negligible. Elimination of the a and b_1A terms produced minimal reductions in R^2, with little change in the ratios b_3/b_2 (Table 17-A). With a and b_1A eliminated, the ratios b_3/b_2 are those given in Table 17-A. The averages of these ratios, $\alpha = b_3/b_2$, were -0.00528 and -0.00870 for the renal resistance and renal conduit segments respectively, for use in the simplified model, Equation 14-2, $y = \beta A^2(1 + \alpha A)$.

At Step Four of curve fitting the change of variable $Z = A^2(1 + \alpha A)$ allows regression estimation of β coefficients in $y = \beta Z$, yielding the β values listed in Table 17-B. Specific variants of Equation 14-2 are plotted in Figure 17-2 to overlie the appropriate plots of observed data. The largest departures of observed data from the fitted curves are generally in groups with as few as 10 or 11 subjects (Table 17-1 gives numbers of cases).

17.6. Steps One to Four, fitting Hy on age

The fittings of Hy to age within tertiles of R_a are presented in Tables 17-A&B. Letting y = Hy, three instances of Equation 14-1 were determined, one each for the Low R_a, Mid R_a, and High R_a groups. The intercept term, a, and the term b_1A were both rejected in every instance. The ratios of b_3/b_2 ranged from -0.01061 to -0.01132 and averaged -0.01092, which rounds to $\alpha = -0.011$ as sufficient in all three instances (Table 17-A). Figure A1 (right panel) plots Equation 14-2 using values of β in Table 17-B for y = Hy in the High R_a and Low R_a groups as solid curves for comparison with observed data plotted as broken lines.

17.7. Curve fitting for four demographic groups

In the curve fitting to data on black and white men and women, every instance of Equation 14-2 succeeded in capturing all statistically significant information of Equation 14-1. These results are documented in Table 17-C, where the outcome of fitting Equation 14-3 in every instance shows lack of statistical significance for all c_n coefficients that were tested. Using R^2 to measure statistical information, the percentage of Equation 14-1 information captured by Equation 14-2 ranged from 96.5 to 99.9 %, and averaged 99.1 %. The condensation of Equation 14-1 into Equation 14-2 requires only three values of α, one for each pathological variable, to be applied alike to all demographic groups. As drawn from Table 17-A these are $\alpha = -0.00528$ when y = R_r and α -0.00870 when y = R_c, and $\alpha = -0.1092$ when y = Hy.

(Technical note: This development was repeated for y = arteriolar hyalinization using the cube root transform, the square root transform, and the raw untransformed measures for Hy. Estimates of α were respectively -0.01128, -0.01117, and -0.01113, averaged over the four demographic groups. These results indicate an absence of any serious distortions due to the use of a square root transform in defining the variable Hy.)

17.8. Regression modeling of the hyaline effect on R_r and R_c

Equation 17-1, with X = Hy, offers another approach to determining the effect of Hy on the progression of renovascular fibroplasia.

$$y = b_1A + b_2A^2 + b_3A^3 + b_4X + b_5XA + b_6XA^2 + b_7XA^3 + a \qquad (17\text{-}1)$$

The regression coefficients of this equation were estimated for y = R_r and y = R_c in black and white men and women. Zero or low statistical significance in all of the eight fittings led to elimination of the terms b_1A, b_4X, b_5XA, b_7XA^3, and a. The simplified equation, $y = b_2A^2 + b_3A^3 + b_6XA^2$, generated R^2 that were reduced by trivial amounts from the total R^2 of Equation 17-1. As in Chapters Fourteen and Fifteen, the ratios of the regression coefficients in the simplified model revealed a simple pattern, which introduces Equation17-2 as an extension of Equation 14-2 to incorporate the hyaline effect.

$$y = \gamma A^2 (1 + \alpha_X X + \alpha_a A) \tag{17-2}$$

Equation 17-2 statistically incorporates virtually all of the information in Equation 17-1 (about 98 % of the variance), and this outcome is demonstrated as follows: Define $Q = A^2(1 + \alpha_X X + \alpha_a A)$ with the properties {1} α_X equals 0.0796 for $y = R_r$ and 0.0221 for $y = R_c$; these are the unweighted averages of the ratios of b_6/b_2 for black and white men and women in Table 17-D. {2} α_a equals -0.00650 for $y = R_r$ and -0.00923 for $y = R_c$; these are the unweighted averages of the ratios of b_3/b_2 in Table 17-6. Equation 17-2 can now be rewritten $y = \gamma Q$, and Equation 17-3 becomes the next step of regression modeling.

$$y = \gamma Q + c_1 A + c_2 A^2 + c_3 A^3 \tag{17-3}$$

The four demographic groups, black and white men and women, supplied data for the two arterial segments in *basal* subjects aged 15-79 years, giving eight versions of the regression Equation 17-3. After entry of γQ into these eight equations, all other terms were rejected as offering no further statistical significance. In these eight instances, the R^2 of Equation17-2 retained an average of 98.5 % of the R^2 in Equation 17-3 (ranging from 94.2 to 99.6 %).

17.9. Estimating parameter γ

For the tabulation of γ coefficients of Equation 17-3 in Table 17-E, the four demographic groups were further subdivided into tertiles of Hy. These subdivisions created groups with small numbers of cases. Nevertheless, after entry of γQ, other terms in Equation 17-3 were rejected as not of significance in all but two instances: The group of 17 black women of High Hy gained added significance by entry of the c_3A^3 term with a positive sign, and the group of 11 white women in the Mid Hy group gained by entry of c_1A with a negative sign. In all but these two instances, therefore, the expression $y = \gamma Q$ is sufficient to capture all of the statistically significant information in Equation 17-1, and the two exceptional instances are of small magnitude sufficient to declare negligible.

The expression $y = \gamma Q$ assumes that the hyaline effect on fibroplasia, i.e. the influence of grouping in accordance with hyaline deposits in arterioles, is the same in all demographic groups and acts uniformly at all ages, and these assumptions cannot be rejected statistically. Furthermore, the γ coefficients do not vary significantly across tertiles of Hy (as shown by ANOVA's that are not tabulated), implying that Equation 17-2 incorporates all of the significant Hy effect across demographic groups of all ages.

17.10. Regression modeling of the R_a effect on Hy

Equation 17-1, with y = Hy and X = R_a, yields the results in Tables 17-D&E. The development imitates the foregoing paragraphs. The outcome supports the conclusion that Equation 17-2 is a satisfactory summation of the way that arteriolar hyalinization evolves with aging under the influence of the grouping effects of R_a in the four demographic groups.

17.11. The universal law of aging in arterial intima

The formula in Equation 14-3, $y = \beta A^2(1 + \alpha A)$, offers the prospect of providing tools for the study of the "aging risk factor" for atherosclerosis. Intimal fibroplasia emerged repeatedly in previous chapters as a profoundly influential precondition for atheroma. Intimal fibroplasia also emerged as a product of aging, whose actions leave a distinctive trail, in the form of adherence to Equation 14-3. We now find in this chapter that arteriolar hyalinization accelerates fibroplasia in total keeping with the formula for aging. This outcome introduces a novel concept: Perhaps aging of arterial intima is not strictly bound to the ticking of atomic clocks, but rather to the ticking of invisible biological clocks whose rates, measured by β, can be made to vary (by means that we can only guess at this time).

Table 17-A. Regression coefficients and R^2 for Low Hy, Mid Hy, and High Hy groups, and for Low R_a, Mid R_a, and High R_a groups in a pool of all *basal* subjects ages 15-79 years.

	Hy Range	Step Two: a Excluded b_2/b_1	b_3/b_1	b_3/b_2	Step Three: b_1A Excluded b_2	b_3/b_2	Step Two R^2	Step Three R^2
$y=R_r$	Low	-0.0773	+0.000604	-0.00781	0.00330	-0.00618	0.6827	0.6770
Renal	Mid	-0.0576	+0.000431	-0.00781	0.00341	-0.00404	0.7477	0.7359
resistance	High	-0.0760	+0.000551	-0.00724	0.00460	-0.00561	0.7916	0.7885
arteries					.Average	-0.00528		
$y=R_c$	Low	+0.0373	-0.000923	-0.00786	0.00746	-0.00855	0.8430	0.8404
Renal	Mid	-0.2283	+0.001941	-0.00850	0.00800	-0.00838	0.8787	0.8684
conduit	High	+0.0713	-0.000667	-0.00936	0.00951	-0.00919	0.8885	0.8876
arteries					.Average	-0.00870		
	R_a Range							
y=Hy	Low	-0.1360	+0.001451	-0.01067	0.00195	-0.01132	0.5367	0.5351
Hyaline	Mid	-0.5063	+0.005312	-0.01049	0.00230	-0.01061	0.6073	0.6072
arterioles	High	+0.2399	-0.002654	-0.01106	0.00292	-0.01082	0.6729	0.6727
					.Average	-0.01092		

At Step One of curve fitting, the regression equation, $y = b_1A + b_2A^2 + b_3A^3 + a$, was tested and the intercept term, a, was found insignificant in all instances. At Step Two, the intercept was forced through the origin; the resulting regression coefficients held the ratios as tabulated. At Step Three, the term b_1A was deleted, leaving $y = b_2A^2 + b_3A^3$, with the tabulated ratios of coefficients. After these three steps, the model carried into the fourth step becomes $y = \beta A^2(1 + \alpha A)$, which imposes a constant value for α in all subsequent models, as estimated by the averages of b_3/b_2 given in this table. Numbers of cases are 138, 116, & 113 in the Low, Mid, and High Hy groups respectively and are 124, 120, and 123 in the Low, Mid, and High R_a groups respectively.

Table 17-B. Step four of regression modeling for Low Hy, Mid Hy, and High Hy groups, and for Low R_a, Mid R_a, and High R_a groups in a pool of all *basal* subjects ages 15-79 years.

	Hy Range	Step Four; constant α				R^2		
		β	c_1	c_2	c_3	Total	Simplified	% Retained
$y = R_r$	Low	0.00253	*	*	*	0.6827	0.6721	98.4
	Mid	0.00316	*	*	*	0.7477	0.7354	98.4
	High	0.00375	*	*	*	0.7916	0.7850	99.2
$y = R_c$	Low	0.00717	*	*	*	0.8430	0.8402	99.7
	Mid	0.00795	*	*	*	0.8787	0.8684	99.9
	High	0.00861	*	*	*	0.8885	0.8856	99.7
	R_a Range							
$y = Hy$	Low	0.00186	*	*	*	0.5367	0.5344	99.6
	Mid	0.00242	*	*	*	0.6073	0.6064	99.9
	High	0.00301	*	*	*	0.6729	0.6724	99.9

In each instance, the equation $y = \beta Z + c_1A + c_2A^2 + c_3A^3$ was determined by regression, where $Z = A^2(1 + \alpha A)$. The tabulated values of β are those when only βZ enters the equation. The symbol "*" indicates regression coefficients that are rejected as not of statistical significance on further entry after βZ. Total R^2 refers to the entire equation including all insignificant terms, the simplified R^2 refers to the equation retaining only βZ, and % retained is their ratio written as a percentage. The parameter α takes the constant values $\alpha = -0.01092$ when $y = Hy$, $\alpha = -0.00528$ when $y = R_r$, and $\alpha = -0.00870$ when $y = R_c$, in keeping with the findings of Table 17-A.

Table 17-C. Regression modeling of Hy and R_a on age for *basal* black and white men and women aged 15-79 years.

		Step Three	Step Four; constant α				R^2		
		b_3/b_2	β	c_1	c_2	c_3	Total	Simplified	% Retained
$y = R_r$	Black men	-0.00379	0.00367	*	*	*	0.7864	0.7755	98.6
	White men	-0.00490	0.00309	*	*	*	0.7460	0.7431	99.6
	Black women	-0.00463	0.00297	*	*	*	0.6794	0.6785	99.9
	White women	-0.00611	0.00251	*	*	*	0.6940	0.6820	98.2
	Average	-0.00486							
$y = R_c$	Black men	-0.00881	0.00836	*	*	*	0.8757	0.8755	99.9
	White men	-0.00860	0.00744	*	*	*	0.8373	0.8356	99.8
	Black women	-0.00849	0.00857	*	*	*	0.8932	0.8916	99.8
	White women	-0.00856	0.00727	*	*	*	0.8689	0.8678	99.9
	Average	-0.00862							
$y = Hy$	Black men	-0.01003	0.00260	*	*	*	0.6703	0.6617	98.7
	White men	-0.01163	0.00272	*	*	*	0.6314	0.6253	99.0
	Black women	-0.01154	0.00178	*	*	*	0.5498	0.5435	98.9
	White women	-0.01094	0.00234	*	*	*	0.5280	0.5093	96.5
	Average	-0.01104							

Footnote as in Table 17-B. In each instance, the ratio of b_3/b_2 obtained at Step Three of model building offers an estimate of the parameter α, as an independent check on values $\alpha = -0.01092$ when $y = Hy$, $\alpha = -0.00528$ when $y = R_r$, and $\alpha = -0.00870$ when $y = R_c$ obtained in Table 17-A and retained also here to calculate β. Numbers of cases are 143, 105, 68, & 52 for black men, white men, black women, and white women respectively.

Table 17-D. R^2 for Equation 17-1 and its simplified form, $y = b_2A^2 + b_3A^3 + b_6XA^2$, and ratios of regression coefficients.

		Ratios of coefficients			R^2		
		b_3/b_2	b_6/b_2	b_6/b_3	Total	Simplified	% retained
$y=R_r$; $X=Hy$	Black men	-0.00647	0.0899	-13.88	0.8171	0.8098	99.1
	White men	-0.00495	0.0234	-0.18	0.7501	0.7472	99.6
	Black women	-0.00663	0.1343	-20.28	0.7524	0.7086	94.2
	White women	-0.00795	0.0706	-8.88	0.7545	0.7388	98.0
	Average	-0.00650	0.0796				
$y=R_c$; $X=Hy$	Black men	-0.01000	0.0318	-3.18	0.8941	0.8895	99.5
	White men	-0.00862	0.0016	-0.18	0.8388	0.8356	99.6
	Black women	-0.00924	0.0373	-4.04	0.9046	0.8989	99.4
	White women	-0.00907	0.0178	-1.96	0.8840	0.8762	99.1
	Average	-0.00923	0.0221				
$y=Hy$; $X = R_a$	Black men	-0.0161	0.02848	-1.77	0.7297	0.7176	98.3
	White men	-0.0126	0.00481	-0.38	0.6352	0.6295	99.1
	Black women	-0.0182	0.03244	-1.78	0.6222	0.5850	94.0
	White women	-0.0136	0.01577	-1.16	0.6543	0.5850	89.4
	Average	-0.0151	0.02038				

Early steps of modeling for Equation 17-1, $y = b_1A + b_2A^2 + b_3A^3 + b_4X + b_5XA + b_6XA^2 + b_7XA^3 + a$, eliminated coefficients a, b_1, b_4, b_5, and b_7. In the remaining expression, $y = b_2A^2 + b_3A^3 + b_6XA^2$, the ratios of coefficients displayed the orderly pattern as tabulated here. The averages of these ratios offer estimates of the Equation 17-2 parameters: For $y = R_r$, $\alpha_h = 0.0796$ and $\alpha_a = -0.00650$. For $y = R_c$, $\alpha_h = 0.0221$ and $\alpha_a = -0.00923$. When $y = Hy$ in the modified Equation 4, $y = b_2A^2 + b_3A^3 + b_6R_aA^2$, $\alpha_R = 0.02038$ and $\alpha_a = -0.0151$.Total R^2 refers to the complete Equation 17-1 and simplified R^2 refers to the reduced expression Equation 17-2; the ratios of these R^2 determines the % retained to represent the percentage of statistical information in Equation 17-1 captured by Equation 17-2.

Table 17-E. Regression modeling of R_r and R_c within Low, Mid, and High ranges of Hy and of Hy within Low, Mid, and High ranges of R_a for *basal* black men (BM), white men (WM), black women (BW), and white women (WW) aged 15-79 years.

		Number of cases	$y = R_r$ in Hy groups, Step Four; constant α					$y = R_c$ in Hy groups, Step Four; constant α					Number of cases	$y =$ Hy in R_a groups, Step Four; constant α				
			γ	c_1	c_2	c_3	% R^2	γ	c_1	c_2	c_3	% R^2		γ	c_1	c_2	c_3	% R^2
BM	Low	51	0.00364	*	*	*	96.5	0.00678	*	*	*	97.3	42	0.00152	*	*	*	93.7
	Mid	51	0.00352	*	*	*	95.1	0.00884	*	*	*	98.6	47	0.00159	*	*	*	92.2
	High	40	0.00356	*	*	*	97.7	0.00826	*	*	*	99.3	53	0.00208	*	*	*	97.0
WM	Low	33	0.00377	*	*	*	97.5	0.00792	*	*	*	99.0	38	0.00197	*	*	*	90.7
	Mid	36	0.00286	*	*	*	98.1	0.00709	*	*	*	98.3	34	0.00210	*	*	*	89.4
	High	36	0.00280	*	*	*	96.5	0.00710	*	*	*	97.4	33	0.00170	*	*	-	88.5
BW	Low	33	0.00298	*	*	*	93.0	0.00886	*	*	*	99.4	21	0.00083	*	*	*	76.9
	Mid	18	0.00305	*	*	*	90.4	0.00782	*	*	*	95.5	24	0.00134	*	*	*	93.1
	High	17	0.00352	*	*	+	81.4	0.00937	*	*	*	98.3	23	0.00159	*	-	*	76.8
WW	Low	21	0.00236	*	*	*	96.7	0.00741	*	*	*	99.9	23	0.00166	*	*	*	87.6
	Mid	11	0.00432	-	*	*	90.5	0.00786	*	*	*	97.5	15	0.00175	*	*	*	86.7
	High	20	0.00240	*	*	*	96.7	0.00826	*	*	*	99.3	14	0.00156	+	*	*	68.4

With $y = R_r$ let $Q = A^2(1 + 0.0796Hy - 0.0065A)$. With $y = R_c$, let $Q = A^2(1 + 0.0221Hy - 0.00923A)$. With $y = Hy$, let $Q = A^2(1 + 0.0204R_a - 0.0151A)$. The equation $y = \gamma Q + c_1A + c_2A^2 + c_3A^3$ takes the tabulated values of γ when only Q enters; "*" indicates that subsequent entry of another term on A is rejected as not statistically significant p>0.05); the "+" or "-" signs indicate the signs taken upon entry of significant terms with added significance. Low, Mid, and High refer to tertiles of Hy when $y = R_r$ or $y = R_c$; Low, Mid, and High refer to tertiles of R_a when $y = Hy$. % R^2 was "Retained" as in Table 17-6.

CHAPTER EIGHTEEN

THE HY EFFECT ON R_a IN WIDELY VARIABLE CIRCUMSTANCES

Abstract. The relationships of Hy and R (whether R_r, R_c, or R_a) to each other and to age are well described by a simple equation that applies under a remarkably wide range of circumstances. The details suggest a causal pattern, $x_1 \rightarrow$ Hy $\rightarrow$ R $\leftarrow x_2$, where the symbols x_1 and x_2 are used to designate two hypothetical separate and independent groups of unknown multifactorial causes.

Intimal fibroplasia in the renal microvasculature can progress rapidly with age in subjects who manifest complete absence of demonstrable hyaline deposits in arterioles [58]. Subjects with severe hyalinization of arterioles, however, typically experience an acceleration of arterial fibroplasia. The magnitude of the Hy effect upon R_a was explored in Chapter Seventeen, with Equation17-2 and Tables 17-1,D&E describing the relationship. This chapter now further explores the relationship by examining differences between diabetics and nondiabetics and between populations around the world.

Diabetics are thought to exceed nondiabetics in the rates of both Hy and R_a. Moreover, substantial differences are also found between the USA and other nations. These widely varying circumstances provide opportunities to challenge Equation 17-2 with tests of its generality. Those tests are pursued in this chapter. With only one exception, Equation 17-2 fully meets these challenges, and it does so using only one variable parameter to encompass the range of diversity afforded by these populations. The one exception, Hy in the Greenland data, could be a consequence of a methodological artifact.

18.1. Review of Chapter Seventeen

Equations 17-1&2 are reproduced here for ease of reference.

$$y = b_1A + b_2A^2 + b_3A^3 + b_4X + b_5XA + b_6XA^2 + b_7XA^3 + a \qquad (17\text{-}1)$$

$$y = \gamma A^2(1 + \alpha_X X - \alpha_a A) \qquad (17\text{-}2)$$

X = Hy when y = R_c or when y = R_r. X = R_a when y = Hy. The substitute values for α_X and α_a are listed in the footnote to Table 18-1.

In Chapter Seventeen, linear regression was used to fit Equation 17-1 to each of the choices for y in four population groups. Stepwise discarding of coefficients led to Equation 17-2 as holding an irreducible minimum number of terms, retaining the most indispensable ones. The values of α_X and α_a were averaged over the separate populations to obtain the fixed quantities to use in all data sets. The change of variable was then employed, with $Q = A^2(1 + \alpha_X X - \alpha_a A)$, so that the linear regression equation

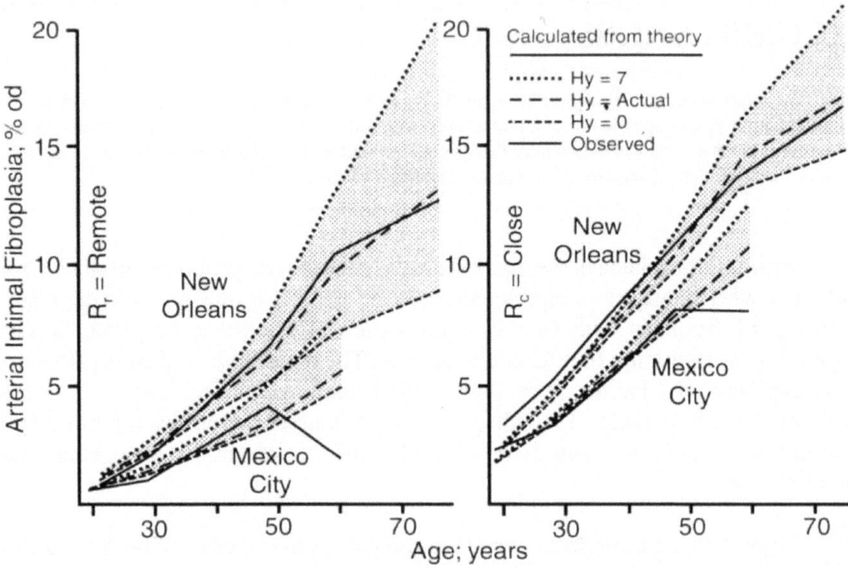

Figure 18-1. Observed means of R_c and R_r within 10-year age groups are joined by angulated solid lines. Means of values calculated from Equation 17-2 are joined by broken lines; these use various arbitrarily chosen substitute values for Hy in the equation. The dashed angulated lines use actually observed means of Hy in each age group as substitute values for calculating the plotted levels of R_c and R_r.

$y = \gamma Q$ could be used to determine the coefficients γ in various populations. Finally, the regression equation $y = \gamma Q + c_1A + c_2A^2 + c_3A^3$ was found to add nothing of statistical significance to the relationship $y = \gamma Q$ in any of the groups of subjects in Tables 17-D or 17-E. Hence, Equation 17-2 can be declared to hold all of the statistically significant information that is in Equation 17-1 for those data sets.

18.2. The sigmoid growth curve

Equation 17-2 is actually a restatement of Equation 14-2 from Chapter Fourteen, the general law of arterial aging (i.e $y = \beta A^2(1 - \alpha A)$). The restatement gives Hy an influence upon the parameter α (with appropriate adjustments of β also). The sigmoid growth curve implied by Equation 17-2 has an inflection point at an age determined by the magnitude of α. As α grows less negative with increasing Hy, the inflection point shifts to an older age. This pattern reflects the fact that Hy governs y ever more strongly with increasing age. Indeed, the effect of Hy is largely confined to the oldest ages.

The pattern is illustrated in Figure 18-1. Diagramed there are versions of Equation 17-2 for men of New Orleans and Mexico City under conditions of X = Hy and y = R_c (right panel) and under conditions of X = Hy and y = R_r (left panel). Three versions of the equation are plotted under each of these conditions, using three arbitrary choices of Hy for substitute values in the equation. Two of the choices are 0 and 7, used to span the usual range of variation for Hy. The third choice is the "actual" magnitude of Hy, which is the observed mean in each of the 10-year age groups.The solid angled lines in Figure 18-1represent the observed means of R_r and R_c, which can be compared with the values "expected" from substituting the "actual" means of Hy. The discrepancies of the two angled lines, one solid and one dashed, serve to measure the disagreements of observed data with the values expected from Equation 17-2. Analysis of variance of these discrepancies confirms the lack of statistical significance of their departures from zero (data not shown).

(Note: In Tables 17-1&2, the means of Hy are nearly constant after age 35 years. Hence, the angled lines of dashes in Figure 18-1 represent cohorts that all happen to share the property of being near the stable "average" that prevails after age 35 years. It can be presumed also that other substitute values of Hy such as 0 or 7 might represent cohorts with those values holding more or less steady over time (cf. Article 14.13). Those with Hy = 0, for instance, may well have held this value nearly constant from ages 35 to 79 years. The same might be said for Hy = 1 or Hy = 2 or other values. At Hy = 7, a strong propensity to die of coronary heart disease might, however, deplete these most severely affected cohorts in the older ages.)

18.3. Comparing New Orleans with Mexico City

Both R_r and R_c are conspicuously lower in the *basal* men of Mexico City than in those of New Orleans (Figure 18-1). Yet, we have reason to think that Hy accelerates R_r and R_c in a similar way in both populations. This is because the same versions of Equation 17-2 are used with identical values of α_X and α_a in both cities. Hy is substantially lower in Mexico (cf. the dashed lines representing mean values are nearer to the lines representing Hy = 0 for Mexico than they are for New Orleans). Yet the accelerating effect of greater Hy in New Orleans can explain only a tiny fraction of the large differences in R_r and R_c between the cities. Rather, it is the variations in parameter γ that account for most of the population differences (Table 18-1).

Table 18-1gives values of γ obtained by regression of $y = \gamma Q$ as explained in Article 18.1. The R^2 of these equations are compared with the R^2 of Equation 17-1 by showing the reductions from 100 % that result from excluding rejected terms. These percentages therefore reflect the amount of statistical information in Equation 17-1 that Equation 17-2 fails to capture in each instance. Very few of these instances encounter statistically significant reductions of R^2, and all are judged to be negligible.

Table 18-1. Parameters for rate of growth on age of renovasculopathies in data sets for all available populations.†

	N Cases	$y = R_r$; %od γ	$y = R_r$; %od % R^2	$y = R_c$; %od γ	$y = R_c$; %od % R^2	y = Hy;$\sqrt{(N/cm^2)}$ γ	y = Hy;$\sqrt{(N/cm^2)}$ % R^2
New Orleans Black							
Men	(177)	0.00360	98.9	0.00804	99.8	0.00191	91.6
Women	(75)	0.00308	94.7	0.00881	99.2	0.00130	84.4
New Orleans White							
Men	(141)	0.00289	97.7	0.00726	99.5	0.00193	91.7
Women	(59)	0.00265	97.8	0.00804	99.8	0.00161	92.0
Mexico							
Men	(111)	0.00189	98.9	0.00580	99.0	0.00148	84.4
Diabetics							
New Orleans	(12)	0.00274	97.9	0.00751	91.0	0.00216	88.0
Mexico	(12)	0.00344	99.3	0.00764	98.7	0.00194	88.5
Tokyo							
Men	(171)	0.00294	98.6	0.00831	99.3	0.00094	99.7
Women	(60)	0.00209	88.0	0.00809	99.4	0.00054	82.3
Greenland							
Men	(41)	0.00266	96.1	0.00751	98.4	.*	
Women	(33)	0.00289	99.2	0.00806	98.1	.*	
Guatemala men	(76)	0.00235	97.4	0.00673	98.7	0.00117	93.7

† The equation $y = b_1A + b_2A^2 + b_3A^3 + b_4X + b_5XA + b_6XA^2 + b_7XA^3$ shrank to $y = b_2A^2 + b_3A^3 + b_6XA^2$ by elimination of insignificant and trivial coefficients, retaining those that best fit all populations and variables with equations of a common form. This in turn simplified to $y = \gamma A^2(1 + \alpha_X X - \alpha_a A)$. As found in Table 17-D, with $y = R_r$ and X = Hy, $\alpha_a = -0.00650$, and $\alpha_X = 0.0796$; with $y = R_c$ and X = Hy, $\alpha_a = -0.00923$, and $\alpha_X = 0.0220$; with y = Hy and $X = R_a$, $\alpha_a = -0.0151$, and $\alpha_{X6} = 0.02038$. Setting $Q = A^2(1 + \alpha_X X - \alpha_a A)$, the values of γ entered in this table were obtained as the coefficients of linear regression for $y = \gamma Q$. The R^2 of $y = \gamma Q$ is reduced from the R^2 of $y = \gamma Q + c_1A + c_2A^2 + c_3A^3 + c_4X$ to the percentages tabulated here.
* = These data call for changing parameters to $\alpha_a = -0.01205$, and $\alpha_X = 0$ because of an unusual behavior of Hy on age (perhaps because of a cohort effect). Thereafter $\gamma = 0.00387$ for men and $\gamma = 0.00267$ for women, reiterating the consistently low values of γ in women when y = Hy.

The magnitude of γ in Mexico City is 53% of that in the Blacks of New Orleans for the variable R_r (0.00189/0.00360 = 0.53) and is 72 % of that in the Blacks of New Orleans for the variable R_c (0.00580/0.00804 = 0.72). It is the tiny branchings of the arterial tree that show the greater difference between populations. It is these tiny branchings that show the greater response to the Hy effect also (cf. wider grey zones for R_r than for R_c in Figure 18-1). Yet there is no apparent connection between these two coincidental relationships.

18.4. Comparing New Orleans with Tokyo, Greenland, and Guatemala City

The estimated values for the parameter γ offer a simple way to compare populations for the rates of progression of microvasculopathies. When $y = R_c$ the γ parameters in men rank the populations in Table 18-1 as New Orleans Black > Tokyo > New Orleans White > Greenland > Guatemala > Mexico and in women New Orleans Black > Greenland > New Orleans White > Tokyo. When $y = R_c$ the γ parameters in men rank the populations Tokyo > New Orleans Black > Greenland > New Orleans White > Guatemala > Mexico and in women New Orleans Black > Tokyo > Greenland > New Orleans White. Many of these differences in rank are small and perhaps negligible; a clear gap occurs only between a high cluster with four members and a low cluster with two members.

When y = Hy, the γ parameters in men rank the populations New Orleans White > New Orleans Black > Mexico > Guatemala > Tokyo and in women New Orleans White > New Orleans Black > Tokyo. A clear gap occurs only between a high cluster of two members and a low cluster of three members. Men are consistently higher than women in all comparisons.

The data on Tokyo are noteworthy in view of the low rates of coronary heart disease that prevail in Japan. Accompanying the low rates of heart disease are low levels of hyaline deposits in arterioles, while fibroplasia of arteries is inappropriately high. These contrasts between populations reiterate the relatively ominous nature of the Hy-linked pathway to coronary atheroma reviewed in Chapter Twenty, a pathway that seems weakly active in the Tokyo data.

(Note: Greenland is the one exceptional data set that rejects the fit of Equation 17-2, and this happens in the one situation when y = Hy. A possible explanation of this is perhaps a cohort effect. Hy forms a peculiar pattern of peaking in the age range 30 - 45 years and declining with age thereafter in these Greenland data. This odd behavior suggests that recent-born cohorts may be more prone than earlier-born cohorts to hyaline deposits in the renal microvasculature. The matter is based on few cases, and remains unresolved at this time.)

18.5. Subjects with diabetic nephropathy

Table 18-2. Means of selected variables for 12 cases in Mexico and 12 cases in New Orleans of diabetic nephropathy.

	Mexico City	New Orleans	t-test	
			t	p
Age; years	57.5	57.8	0.00	0.95
Renofibroplasia				
Close; %od	13.9	13.6	0.01	0.93
Remote; %od	10.9	9.9	0.76	0.39
Renohyalin; $\sqrt{(N/cm^2)}$	3.9	4.1	0.07	0.68
Glomerulosclerosis				
Nodular; % glomeruli	7.6	4.4	1.28	0.27
Diffuse Type A; % tissue	89.8	85.1	1.62	0.12
Diffuse Type B; % tissue	93.5	90.9	1.34	0.20
Obsolete glomeruli; N/sample	42.4	29.8	0.81	0.38

New Orleans and Mexico City each supplied 12 autopsies of subjects with diabetic nephropathy who died of diabetic complications (cf. Article 18.9). Variables observed in these 24 subjects are described in Table 18-A. The 12 subjects with diabetic nephropathy in Mexico City do not differ significantly from the 12 matched subjects in New Orleans in mean values for any of the eight assessed variables (Table 18-2).

Figure 18-2 is a scatter plot of Hy versus age in the 24 diabetics. Into this diagram, lines plotting the previously determined *basal* groups are also introduced. Points representing diabetic subjects tend to fall above the *basal* lines of both cities, but significantly so only for Mexico City reference line. Of the 24 diabetics, 22 fall above the *basal* line of Mexico City, a strongly significant difference from the expected value of 12 subjects. Of the 24 diabetics, 16 fall above the *basal* line of New Orleans; by χ^2 test, this is not a significant excess above the expected number of 12 subjects ($0.1 < p < 0.25$).

Figure 18-3 is a similar plot for R_c, and lines plotting *basal* groups are copied from Figure 18-1 for comparison as solid and dashed angulated lines. The diabetics reveal a strong tendency to fall above the line for *basal* subjects of Mexico, but do not differ measurably from the *basal* subjects of New Orleans. Of the 24 diabetics, 21 fall above the line for *basal* subjects of Mexico, a strongly significant difference from the expected value of 12 subjects. The line for *basal* subjects of New Orleans divides the diabetic cases into 11 above and 13 below, essentially a random division.

18.6. The γ parameters for diabetics

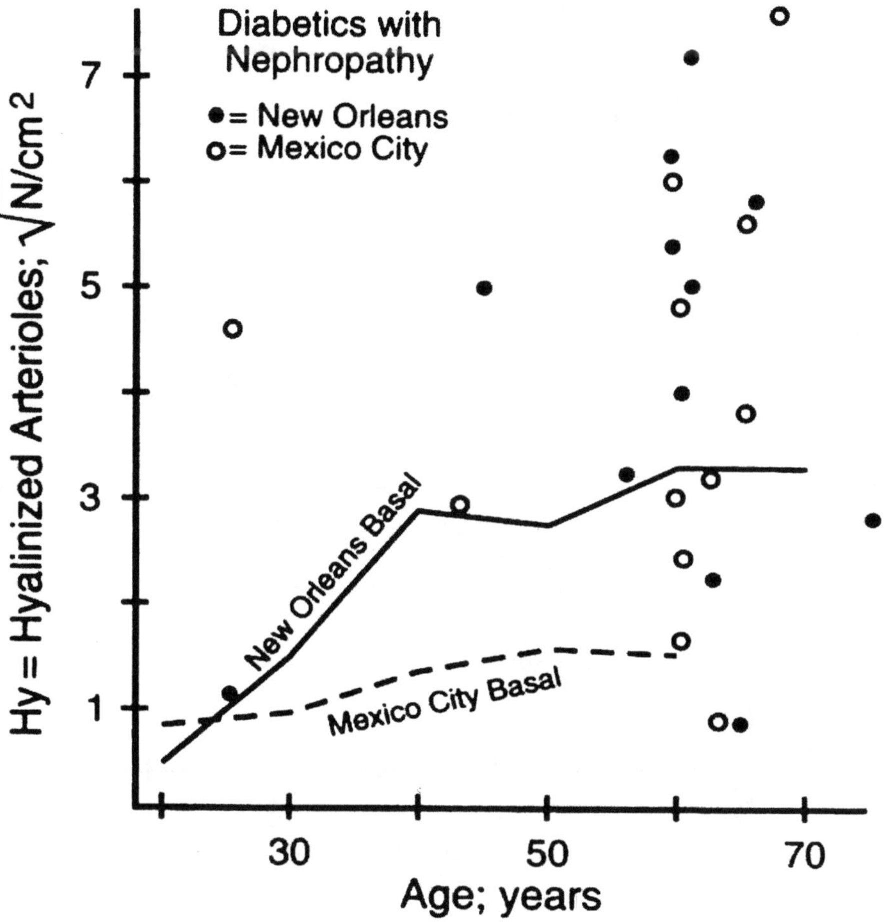

Figure 18-2. Spots and circles plot the extent of arteriolar hyalinization against age for 24 diabetics with nephropathy in two populations. Angulated lines joining means of 10 year age groups represent the average degrees of hyalinization seen in the basal (cardiovascular disease free) nondiabetics of the two populations.

These parameters are given in Table 18-1. When comparing these with other entries in the table, we must recall that the diabetics comprise 8 men and 16 women, so that the comparison data on women are of greater relevance. When y = Hy, $\gamma = 0.00205$ for an average of the 24 diabetics compared with $\gamma = 0.00161$ for an appropriately weighted average of *basal* black and white men and women of New Orleans and $\gamma = 0.00148$ in

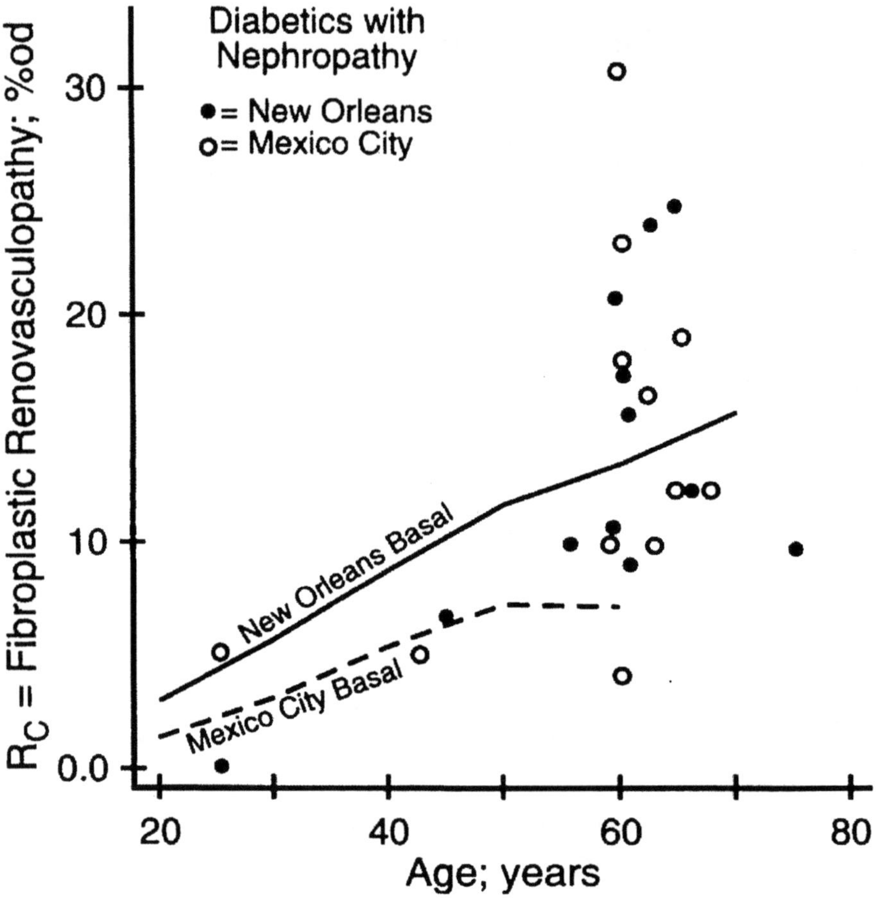

Figure 18-3. Mean severity of arterial intimal fibroplasia is plotted as in Figure 18-2.

basal men of Mexico. Diabetics have accelerated the progression of Hy in both populations, but conspicuously so only in Mexico City. This outcome is a restatement of the results given in Article 18.5.

When $y = R_c$, $\gamma = 0.00757$ in diabetics compared with $\gamma = 0.00817$ in a weighted average of *basal* New Orleans subjects and $\gamma = 0.00580$ in *basal* men of Mexico. Diabetes conspicuously accelerated R_c in Mexico but not at all in New Orleans. A similar outcome follows for $y = R_r$ (data not shown).

The paradoxical outcomes raise an intriguing possibility: Could it be that the rates of progression for microvascular fibroplasia have limited ranges of what is biologically possible? Perhaps the rates in New Orleans, Tokyo, and Greenland are at the maximum of what is possible, and diabetes is unable to force this any further. Perhaps diabetes can reveal its impact only when acting upon the much lower baseline seen in Mexico.

(Note: The findings in diabetics are unable to test the goodness of fit of Equation 17-2. Numbers of cases are too few and with a narrowly limited range of ages. Such testing remains to be done. The results discussed in this section of text make the assumption that the equation is appropriate. These results must therefore be viewed with great caution. That caution is not needed however when reading Figures 18-2&3.)

18.7. A speculative synthesis

As a general principle, the persistent, reproducible correlation of two variables with each other often points to the existence of a causal association between them. Either one causes the other, the other causes the one, or some third variable causes both of them. In the present setting, the structure of the data sets favors the following causal scheme: $x_1 \rightarrow Hy \rightarrow R \leftarrow x_2$. This layout postulates two separate but perhaps overlapping unknown causes for the conditions Hy and R (whether R_r, R_c, or R_a), and then presumes Hy acts to mark the presence of a circumstance of accelerated progression of R. This layout is a logical deduction growing from the wide generality of Equation 17-2. The separation of the unknown causes, x_1 and x_2, is supported by the contrast between Tokyo and other cities. A single version of Equation 17-2 governs the progression of R in both Tokyo and New Orleans in spite of a vast contrast in the progression of Hy. The same single version of Equation 17-2 governs the progression of Hy in both Tokyo and Mexico City in spite of a vast contrast in the progression of R. On the other hand, the overlap of the unknown causes, x_1 and x_2, is suggested by diabetics, because diabetes accelerated both Hy and R in the Mexican data (raising the γ parameters of diabetics in Equation 17-2).

The causal direction of Hy $\rightarrow$ R is supported by the widening of the grey zones with age in Figure 18-1. Hy attains a stable level after age 35 years, yet continues to exert an ever expanding effect upon R. The reverse direction, Hy $\leftarrow$ R is unreasonable, because the presumed consequence, Hy, fails to increase with age, while the presumed cause, R, rises unremittingly into old age.

The components of the unknown causes x_1 and x_2 are apt to be multiple. It seems likely that these components do *not* include high blood pressure. The body parts with the highest blood pressure, the feet, are among the most resistant to both Hy and R, while the body part with the lowest blood pressure, the brain, is among the most prone to Hy and R (while sparing the skull and its muscle, fat, and skin just as in the feet). The most prone body parts are the solid abdominal viscera, and this propensity can in no way be linked to blood pressure. Experimental production of high blood pressure in animals fails to produce Hy, which is a feature found in the human, affecting other animal species under only the most extreme of experimental circumstances. The anatomic structure of R produced in hypertensive animals fails to imitate the natural condition of man in a number of respects. The weight of evidence discourages the view of benign essential hypertension as a vicious cycle. High blood pressure typically persists for decades without any apparent acceleration of further hypertension or of renal microvasculopathy (although renal function can deteriorate for reasons unrelated to arterial conditions). The vicious cycle of pressure causing vasculopathy causing higher pressure and so on is a description of "malignant hypertension". If such a cycle prevailed generally, then all hypertension ought to be malignant, something wholly contrary to observation. When seeking the unknown causes, x_1 and x_2, we should look elsewhere than high blood pressure.

18.8. Populations

18.8.1. Mexico City

A series of forensic autopsies in men who died of violence at ages 15-59 years included only two subjects over age 54 years. Paraffin-fixed samples of renal cortex were transported to New Orleans for paraffin embedding, sectioning, and staining with PAS-Alcian blue as for all other population groups. The slides were interspersed in batches among the slides in the ongoing series of cases in New Orleans [57].

18.8.2. Tokyo

The Toho University School of Medicine supplied renal tissues from 270 autopsies [58]. These tissues were from men and women aged 15-92 years dying of various causes, mostly cancers, during 1986-87 or from 1995-96. A *basal* group was constructed by excluding all instances of cardiovascular deaths (CVD, 15 cases) and chronic renal disease (15 cases), retaining 240 cases.

18.8.3. Greenland

Routine autopsies were made whenever possible in the two municipalities Nuuk and Ilulissat [70]. Both towns are placed on the Greenland West Coast respectively approximately 250 miles south and north of the Polar Circle. Nuuk has 13,000 citizens; Ilulissat 4,000. With each death of a native Greenlander, permission for autopsy was sought from the relatives. In each permitted case, autopsy was carried out by a local physician trained in the procedure. The autopsy rate was 27% in the period of specimen collection. Of the 88 cases, 63 were residents of Nuuk, 19 of Ilulissat, and 16 of other locations. Cerebrovascular disease of indeterminate type was given as the cause of death in seven cases, coronary heart disease in five, gangrene of the leg in one, and ruptured aortic aneurysm in one. After removal of these 14 cases, the remaining 74 cases form the *basal* group, retaining only those kinds of diseases that have no known relationship to atherosclerosis or hypertension.

18.8.4. Guatemala

Samples of kidney tissue were gathered as part of the protocol for the study of atherosclerosis [68]. In 1973, specimens were obtained under the supervision of personnel from New Orleans, from medicolegal autopsies in Guatemala City. The protocol called for 20 men in each 10-year age group, 15-24, 25-34, 35-44, and 45-54 years. Formalin fixed samples of renal cortex were transported to New Orleans for embedding and sectioning in paraffin, staining, and evaluation along with specimens from New Orleans. Data were successfully obtained in 77 cases.

18.9. Diabetics

Unidad de Patologia del Hospital General de Mexico y de la Facultad de Medicina, Universidad Nacional Autonoma de Mexico, in Mexico City, supplied 12 diabetic cases. The autopsy files of 1993 were screened for the first 12 suitable subjects with diabetic nephropathy. These were quickly assembled within the first few months of the year, and comprised four men and eight women. Paraffin blocks of renal tissue were retrieved and prepared as slides to be transported to New Orleans for later staining and evaluation. Age distribution is shown in scatter graphs. The autopsy files of Charity Hospital in New Orleans for the years 1990-1994 (blacks only) were reviewed to obtain instances of diabetic nephropathy matched to those from Mexico by age and sex. Of the 12 Mexico cases, cause of death was given as diabetic nephropathy in six, coronary heart disease in three, infections in two, and hepatic cirrhosis in one. Of the 12 New Orleans cases, cause of death was given as diabetic nephropathy in five, coronary heart disease in five, and infections in two.

18.10. Diabetic nephropathy

Simple techniques were applied to assess whether diabetic nephropathy in New Orleans specimens was of similar degree to that in Mexican specimens. The first 40 suitable glomeruli encountered in each slide were selected for evaluation. Obsolete glomeruli were excluded as not suitable (Figure 18-4B), and a record was kept of their numbers (N excluded per sample of 40 suitable glomeruli). Also recorded was the percentage of suitable glomeruli having nodular intercapillary glomerulosclerosis (KW = Kimmelstiel-Wilson nodules, Figure 18-4A). Diffuse intercapillary glomerulosclerosis was assessed separately in the glomeruli with and without KW nodules; this was done by a point counting method. A grid in the eyepiece of the microscope was placed over the glomerulus, tangential to it at the 9 O'clock and 12 O'clock positions (Figure 18-5). The points marked by crossing grid lines were counted as falling upon solid tissue (capillary wall or mesangium indiscriminately) or upon hollow space within the outline

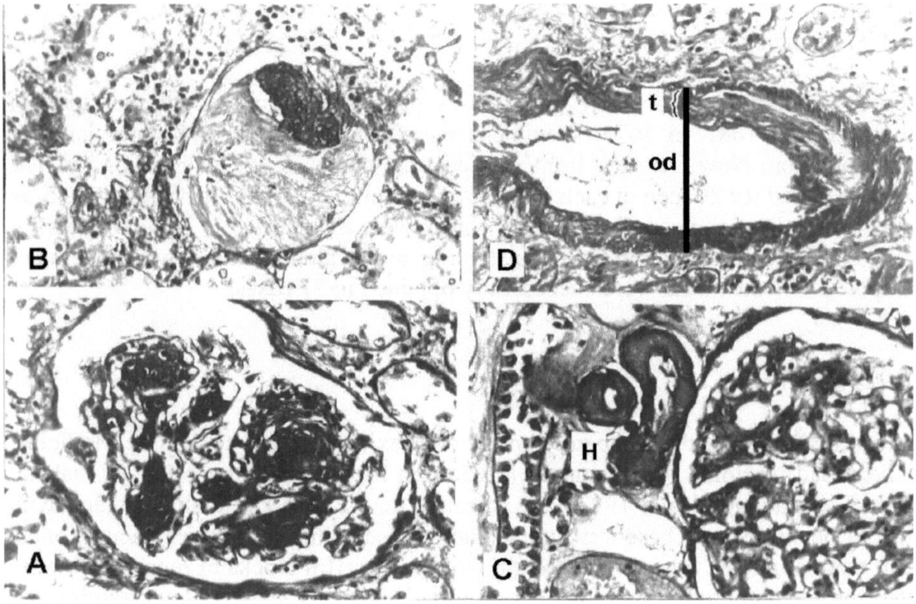

Figure 18-4. A: Nodular intercapillary glomerulosclerosis (Kimmelstiel-Wilson = KW) lesions are illustrated. B: An obsolete glomerulus is shown. C: A helical arteriole, cut across twice, is intensely hyalinized (H), in company with a glomerulus mildly affected by diffuse intercapillary glomerulosclerosis. D: Fibroplastic intima, t, is related to the outer diameter, od, of an interlobular artery to give R = 100t/od, a measure of fibroplastic renovasculopathy. PAS, line od is of length 120 μm , magnification alike in all frames.

of the capillary tuft (capillary lumen or urinary space indiscriminately). The percentage of points falling upon solid tissue was averaged over the two types of glomerulus (S_A or S_B, having or not having KW nodules respectively). A series of kidneys, chosen to represent nondiabetics, was constructed by using the first five in a batch of forensic cases with normal histology. In these, average percentage of points falling upon tissue was 46.3 % ± 8.5 %. This method is crude in comparison with similar methods applied to electron micrographs, but adequately serves the limited uses required by this study. Arterial fibroplasia (Figure 18-4D) and hyalinized arterioles (Figure 18-4C) and were assessed as in Chapters Twelve and Seventeen respectively.

18.11. Description of variables

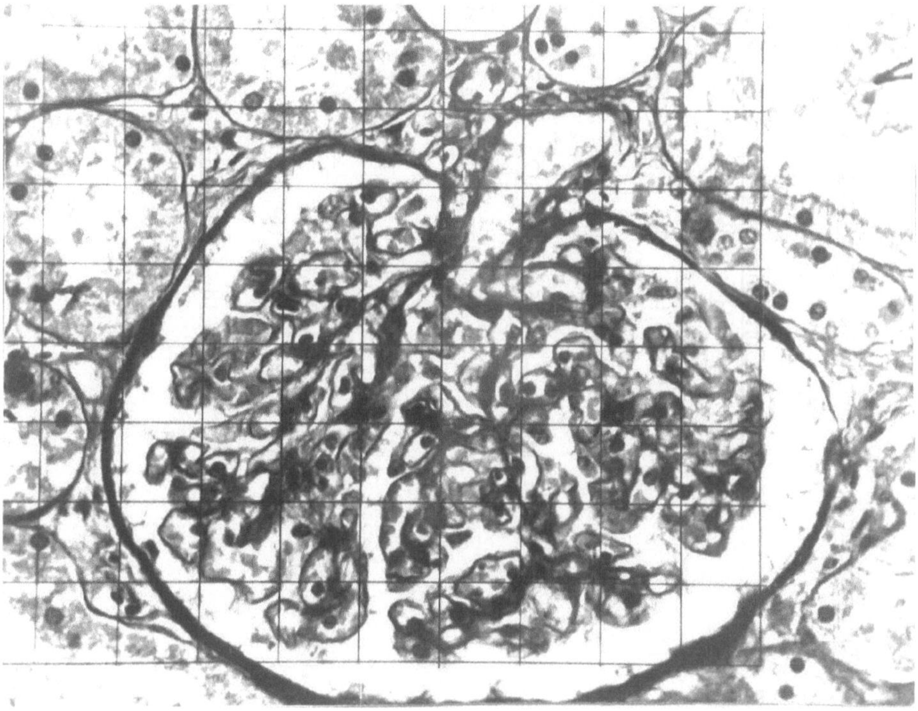

Figure 18-5. A normal glomerulus illustrates the method of point counting used to assess the percentage of glomerular tuft composed of solid tissue. Grid lines are separated by 25 μm, PAS.

Table 18-A presents a list of the variables used in this chapter, together with descriptive information about them. This system applies to the total set of 24 diabetic subjects, combining men and women of Mexico and New Orleans. These statistics provide the information needed for multivariate analyses including correlation and regression.

Table 18-A. Descriptive characteristics of selected variables, including correlation coefficients, raw (upper triangle) and age adjusted (lower triangle) in 24 examples of diabetic nephropathy.

Variable	Symbol	Units	Mean	SD	Correlation coefficients						
					R_c	R_r	H	KW	S_A	S_B	G_O
Age	A	Years	57.7	11.7	.51**	.52**	.10	.00	.05	.59**	.25
Renofibroplasia											
Close	R_c	%od	13.7	7.4	.---	.77**	.01	-.12	-.05	.21	.44*
Remote	R_r	%od	10.0	5.4	.69**	.---	.35	-.10	.23	.35	.31
Renohyalin	H	$\sqrt{(N/cm^2)}$	1.9	4.0	-.09	.31	.---	.37	.18	.41*	.18
Glomerulosclerosis											
Nodular	KW	%	6.0	6.9	-.13	-.11	.38	.---	.28	.31	.19
Diffuse Type A	S_A	%	92.4	3.8	-.09	.23	.17	.28	.---	.69**	.34
Diffuse Type B	S_B	%	87.5	8.9	-.14	.06	.39*	.39*	.82**	.---	.17
Obsolete glom	G_o	N/sample	36.1	34.0	.38	.22	.14	.20	.33	.34	.---

* = significantly different from zero at $p < 0.05$, ** at $p < 0.01$

CHAPTER NINETEEN

HYALINIZED RENAL ARTERIOLES AND THE MALENESS CORONARY RISK FACTOR

Abstract. The two aspects of fibroplastic coronary intimal thickening, those linked to C_T and F_C, relate asymmetrically to the renovasculopathies, R_a and Hy. R_a associates chiefly with the C_T-linked aspect and Hy with the F_C-linked aspect of coronary fibroplasia. Since the F_C-linked aspect is the more ominous type, it appears that Hy is the more ominous kind of microvasculopathy. The association of Hy with F_C seems to happen exclusively in men, and is statistically sufficient to "explain" the maleness risk factor for coronary atheroma.

Two aspects of fibroplastic coronary intimal thickening emerged from inquiries in the foregoing chapters, an aspect attributed to SMC's producing excessively large realms for themselves to occupy, measured by F_C, and an aspect attributed to excessively numerous SMC's each producing a proportionate amount of fibroplasia, measured by C_T. The data examined in this chapter find these two aspects to correlate with the two types of renovasculopathy, hyaline, Hy, and fibroplastic, R_a, with important asymmetries. Those asymmetries fall upon the coronary arteries and renal arterioles of women in ways that differ from those in men. These findings offer useful insights into the "maleness" risk factor that promotes early formation of atheroma in the coronary arteries of men.

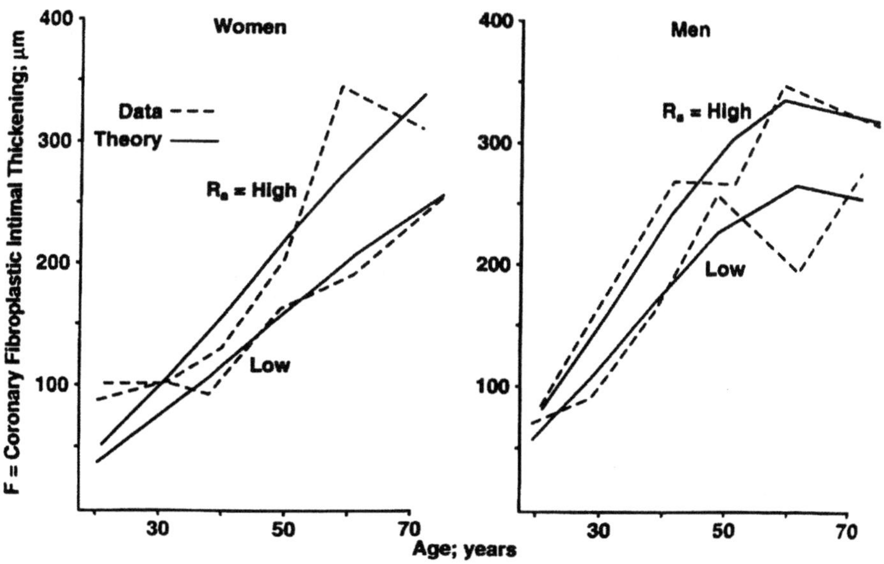

Figure 19-1. The variable F represents coronary fibroplastic intimal thickenings as illustrated in Figure 1-1B and cover photograph 3a. Dashed lines are observed data reproduced from Table 19-1 and solid curves represent Equation 14-2 with appropriate choices for α and β.

Table 19-1. Means of coronary fibroplasia measures for Low R_a, Mid R_a, and High R_a tertiles of renal arterial fibroplasia in 10-year age groups; *basal* pool of black and white subjects.

Age; years	F; µm			F_c µm/SMC			N cases		
	Low	Mid	High	Low	Mid	High	Low	Mid	High
--- MEN ---									
15-24	72	96	84	4.9	5.4	5.3	15	13	15
25-34	94	121	165	5.4	6.4	6.2	19	15	16
35-44	165	212	270	7.6	7.9	9.5	19	25	24
45-54	260	256	268	9.6	9.4	9.5	16	15	20
55-64	197	234	352	10.5	10.3	14.6	5	7	7
65-79	290	265	317	11.2	11.5	11.7	6	5	5
Unweighted Mean†	180	198	243	8.2	8.5	9.5			
	A	A	B	A	A.B	.B			

ANOVA	F	(p)	F	(p)
Age	21.6	0.00	20.4	0.00
Renofibroplasia	5.8	0.00	2.5	0.08
Interactions	1.0	0.48	0.8	0.60

Age; years	F; µm			F_c µm/SMC			N cases		
	Low	Mid	High	Low	Mid	High	Low	Mid	High
--- WOMEN ---									
15-24	89	70	104	5.5	4.6	5.3	5	6	7
25-34	104	95	110	4.8	5.7	5.5	7	11	9
35-44	95	110	128	6.1	6.7	5.9	13	6	9
45-54	166	167	206	7.1	7.4	7.7	9	9	4
55-64	194	128	349	8.1	8.1	11.5	5	3	4
65-79	261	267	315	12.7	12.4	13.9	5	4	4
Unweighted Mean†	151	140	202	7.4	7.5	8.3			
	A	A	B	A	A	A			

ANOVA	F	(p)	F	(p)
Age	21.7	0.00	22.4	0.00
Renofibroplasia	8.4	0.00	1.3	0.27
Interactions	1.6	0.12	0.6	0.80

† Unweighted means for each variable that fail to share a symbol A or B differ significantly at $p < 0.05$.

19.1. Coronary total fibroplasia, F, in tertiles of R_a

Table 19-1 presents mean fibroplastic intimal thickness of the coronary artery within tertiles of R_a constructed as explained in Chapter Seventeen, Article 17.3. These data are plotted in Figure 19-1 as dashed curves, omitting the intermediate Mid R_a groups to avoid cluttering the chart. Solid curves represent Equation 14-2, $y = \beta A^2(1 + \alpha A)$, with $y = F$ and α derived as detailed in Tables 19-A&B appended to the end of this chapter. The steps of curve fitting elaborated in earlier chapters are expanded here by examining women separately from men, with the important outcome that $\alpha = -0.01031$ for men differs conspicuously from $\alpha = -0.00726$ for women. These values correspond to sigmoid curves with inflection points at age 46 years for women and 32 years for men.

In men, F rises swiftly from ages 20 to 40 years, and increasingly more slowly thereafter. In women, F differs little from that in men at age 20 years, but sharply lags the steep early rise seen in men. Yet after age 50 years, the women continue to expand the intimal thickness measured by F, rather than slackening as the men do into old age. By age 70 years the sex difference reverses.

The results shown in Tables 19-A&B imply that the α parameters assigned to the coronary arteries of men and women, although different from each other, can each be treated as constant across R_a groupings and across age groupings. These assumptions are incorporated into Equation 14-2 (Chapter Fourteen), and they cannot be rejected statistically. Hence the magnitude of F in high R_a groups at each age differs from that in low R_a groups solely as measured by β as given in Table 19-B. F increases 28 % faster in high R_a men than in low R_a men ($\beta = 0.251$ and 0.196 respectively) and 37 % faster in high R_a women than in low R_a women ($\beta = 0.141$ and 0.103 respectively). These results serve to open interesting biological speculations. Taking Equation 14-2 as a characteristic insignia of the "aging risk factor", that insignia shows no indication of modifications from one age group to the next. The course through life, whether swift or laggard, seems set at the onset of adult ages, without later course alterations.

19.2. Coronary total fibroplasia, F, in tertiles of Hy

Table 19-2 presents mean fibroplastic intimal thickness of the coronary artery within tertiles of Hy, constructed as explained in Chapter Seventeen, Article 17.1. The table portrays a conspicuous difference of women from men. The grouping by hyaline deposits in renal arterioles has a substantial impact on F in men, but none at all in women. This stands in striking contrast to Table 19-1, where F in women reveals as great a response to R_a as in men.

Figure 19-2 expands upon this important sex difference. The diagonal lines in the graphs represent Equation 19-1, which is the discriminant function that optimally distinguishes High Hy subjects from Low Hy subjects, where $F_c = F/C_T$ and $W_H = 0$ for

Table 19-2. Means of coronary fibroplasia measures for Low Hy, Mid Hy, and High Hy tertiles of renal arteriolar hyalinization in 10-year age groups; *basal* pool of black and white.

Age;	F; μm			F_c μm/SMC			N cases		
years	Low	Mid	High	Low	Mid	High	Low	Mid	High
				--- MEN ---					
15-24	89	67	84	5.1	4.5	6.3	27	10	6
25-34	116	136	125	5.6	7.8	5.5	19	18	13
35-44	158	211	267	6.8	7.8	9.8	15	28	25
45-54	241	254	285	9.0	9.4	10.2	15	17	19
55-64	180	360	245	7.4	16.6	10.8	4	6	9
65-79	245	283	336	10.8	10.4	13.4	4	8	5
Unweighted Mean†	171	218	224	7.5	9.2	9.4			
	A	B	B	A	B	B			

ANOVA	F	(p)	F	(p)
Age	18.6	0.00	18.8	0.00
Renohyaline	3.8	0.02	5.7	0.00
Interactions	1.4	0.16	3.1	0.00

Age; years	F Low	F Mid	F High	F_c Low	F_c Mid	F_c High	N Low	N Mid	N High
				--- WOMEN ---					
15-24	78	101	108	5.2	4.7	5.1	11	3	4
25-34	80	91	124	4.6	5.3	5.7	6	10	11
35-44	113	83	115	5.7	6.7	7.4	15	5	8
45-54	172	152	191	6.8	7.1	7.7	9	6	7
55-64	230	256	176	9.8	8.3	9.1	6	4	2
65-79	296	257	262	12.3	12.1	13.1	7	1	5
Unweighted Mean†	162	156	162	7.5	7.2	8.1			
	A	A	A	A	A	A			

ANOVA	F	(p)	F	(p)
Age	11.9	0.00	12.2	0.00
Renohyaline	0.0	0.95	.0	0.39
Interactions	0.6	0.84	0.4	0.93

† Unweighted means for each variable that fail to share a symbol A or B differ significantly at $p < 0.05$.

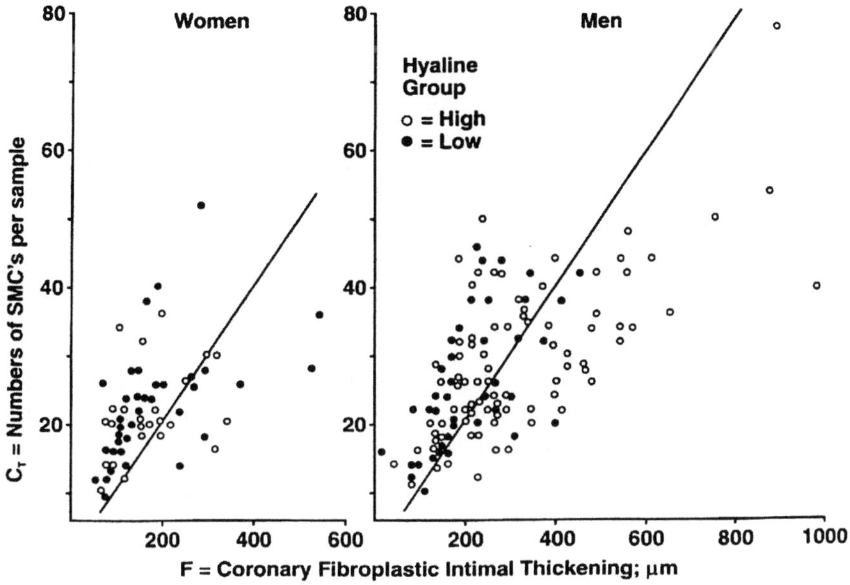

Figure 19-2. Each symbol represents the means of the two variables, F and C_T, observed in an autopsy. SMC's are enumerated as illustrated in Figures 1-1A&B (Chapter One). The solid diagonal line represents Equation 19-1, the discriminant function that optimally separates circles from dots as fit to the men and copied into the plot for women; it is chosen to pass through the origin of the graph and to illustrate C_T in direct proportion to F, but is not the regression line fit to the data. Men and women aged 35-79 years with all causes of death.

optimal discrimination. Symbols are detailed in Chapter One, Article 1.7.

$$W_H = 0.23F_C \quad (T^2 = 0.39 \quad \% = 58 \ c = -10) \tag{19-1}$$

Equation 19-1 was derived from data on men aged 35-79 years and its plot was later copied for comparison into the scatter plot for women. The graph of Equation 19-1 (after conversion of the standardized coefficient to its natural form through division by the standard deviation from Table 1-1a) passes through the origin and by definition must depict F in direct proportion to C_T.

In the plot for men, many symbols fall rightward of the solid diagonal but few symbols do so in the plot for women. It is the High Hy and not the Low Hy groups that account for this difference between men and women. In the Low Hy groups (the dots), 8 of 36 = 22 % of women and 12 of 41 = 29 % of men fall rightward from the diagonal. In the High Hy groups (the circles), 5 of 23 = 22 % of women and 43 of 83 = 52 % of men fall rightward of the diagonal. Subjects falling rightward of the diagonal are those with excessive production of fibroplastic matrix per SMC in the coronary intima. Such subjects often have severely hyalinized renal arterioles, and they usually are men.

A greater proportion of men than of women occupy the High Hy categories. But this is not the conspicuous sex difference in Figure 19-2. Rather, what stands out is the shifting rightward across the diagonal by most of the High Hy men, but the failure of most High Hy women to make such a shift. The coronary intima of men seems to respond briskly to an effect related to Hy grouping, while the coronaries of the women resist that influence.

19.3. Coronary fibroplasia per SMC in relation to Hy

Of the two aspects of coronary fibroplasia, it is the F_C-linked and not the C_T-linked fibroplastic thickening of coronary intima that bears the stronger association with the third variable, Hy (Figure 19-3, sloping line plots Equation 19-2).

$$W_H = 0.63F_C + 0.32C_T \quad (T^2 = 0.49 \;\; \% = 61 \;\; c=-2.26) \qquad (19\text{-}2)$$

Equation 19-2 expands Equation 19-1 by including the C_T term. The plot of this equation is nearly vertical. Intimal thickening put into place by many SMC's, reflected in the variable C_T, has some association with High Hy groups, but excessive thickening produced by each SMC, reflected in F_C, has a more prominent association (by about 2:1 $\approx$0.63:0.32).

Equations 4-4,5,&6 in Chapter Four found atheroma and fatty streak elements to relate better to F_C-linked than to C_T-linked fibroplasia. Equation 19-2 now reiterates this pattern. This coincidence tends to associate severe arteriolar hyalinization, Hy, with the more ominous form of coronary fibroplasia, F_C; both conditions often coexist in the same autopsy.

19.4. Coronary fibroplasia per SMC in relation to R_a

Equation 19-3 is the discriminant function that optimally separates men of High R_a from men of Low R_a groupings.

$$W_R = 0.15F_C + 0.32C_T \quad (T^2 = 0.13 \;\; \% = 52 \;\; c=-1.16) \qquad (19\text{-}3)$$

The relative strengths of the two coefficients, 0.15:0.32 compared with 0.63:0.32, indicate a reversal of relationships from Equation 19-2 to Equation 19-3, with the C_T term coming to overwhelm the F_C term (the F_C term is not statistically significant, $p > 0.1$). When plotted in Figure 19-4, Equation 19-3 forms a nearly horizontal line, in contrast with the nearly vertical line in Figure 19-3. The two kinds of

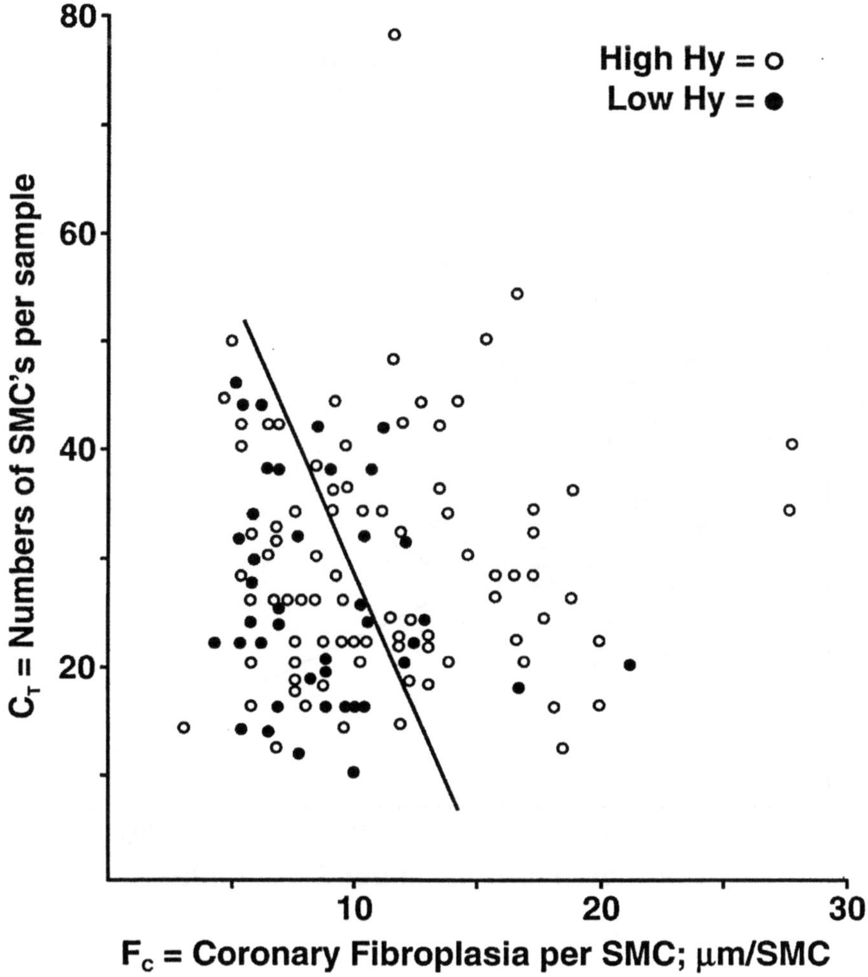

Figure 19-3. As in Figure 19-2, replacing F by F_C on the horizontal, men only. Nearly vertical line represents Equation 19-2.

renovasculopathies reveal a clear asymmetry. R_a relates best to C_T, the less ominous aspect of fibroplasia, while Hy relates best to F_C, the more ominous aspect.

This outcome is strikingly different in women, as will be examined next in Chapter Twenty.

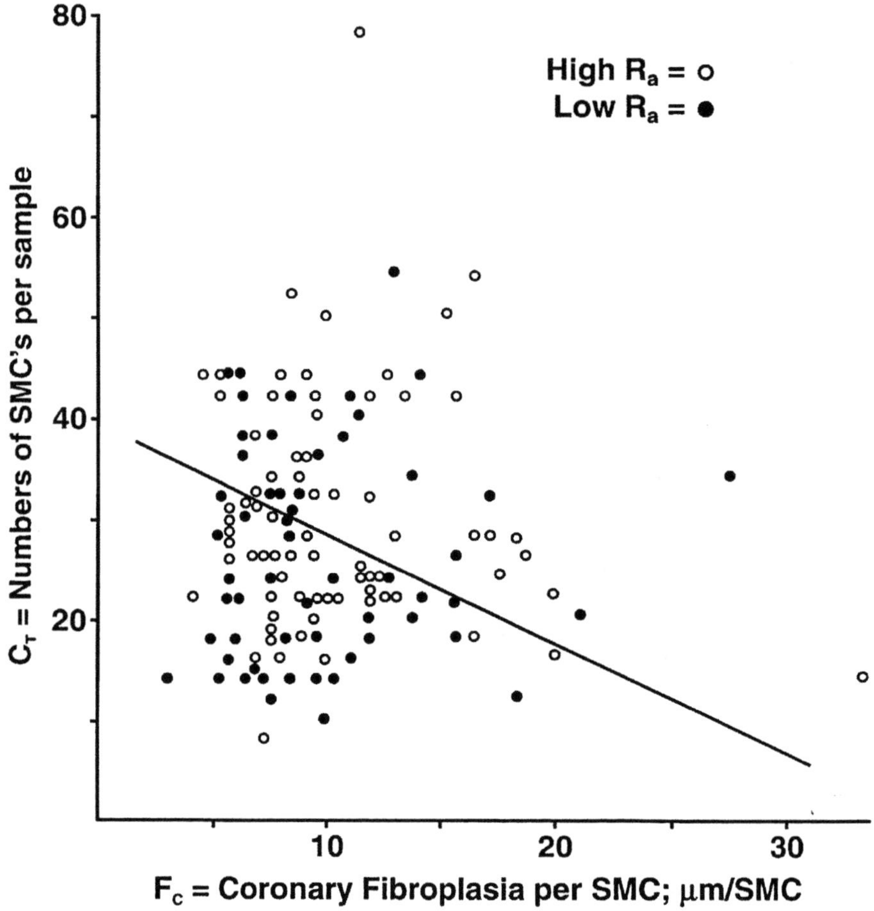

Figure 19-4. As in Figure 19-3, but with R_a groups rather than Hy groups distinguished by circles and dots. Nearly horizontal line represents Equation 19-3.

19.5. Appendix tables

The tables appended here present the steps of curve fitting to the data sets used for Tables 19-1&2. These appended tables document the important conclusion that the parameter α can be treated as constant across groupings based upon R_a. This consistently recurring outcome supports the proposal of Equation 14-2 as a universal law of arterial aging which can serve as an insignia for recognizing the workings of the "aging risk factor". A similar analysis for grouping on Hy is deferred to the next chapter.

Table 19-A. Steps Two and Three of regression modeling of Fibroplastic Coronary Intimal Thickenings for Low R_a, Mid R_a, and High R_a groups in pools of *basal* men and women aged 15-79 years.

R_a Range	Step Two a excluded b_2/b_1	Step Two a excluded b_3/b_1	Step Two a excluded b_3/b_2	Step Three b_1A excluded b_2	Step Three b_1A excluded b_3/b_2	Step Two b_1A Included R^2	Step Three b_1A Excluded R^2
--- MEN ---							
Low	0.0659	-0.000658	-0.00999	0.1832	-0.00974	0.7612	0.7597
Mid	0.1185	-0.001343	-0.01133	0.2264	-0.01079	0.8460	0.8453
High	0.0850	-0.000936	-0.01101	0.2538	-0.01041	0.7906	0.7895
				Average	-0.01031		
--- WOMEN ---							
Low	-0.0168	-0.000181	-0.01076	0.1331	-0.00562	0.8782	0.8633
Mid	-0.0093	0.000116	-0.01247	0.1112	-0.00789	0.8856	0.8743
High	0.1741	-0.001408	-0.00826	0.1576	-0.00827	0.8426	0.8402
				Average	-0.00726		

At Step One of curve fitting, the regression equation, $y = b_1A + b_2A^2 + b_3A^3 + a$, was tested with $y = F$ and the intercept term , a, was found insignificant in all instances. At Step Two, the intercept was forced through the origin; the resulting regression coefficients held the ratios as tabulated. At Step Three, the term b_1A was deleted, leaving $y = b_2A^2 + b_3A^3$,with the tabulated ratios of coefficients. After these three steps, the model carried into the fourth step becomes $y = \beta A^2(1 + \alpha A)$, which imposes the constant ratio of α = the average b_3/b_2 in all subsequent models, as estimated by the averages given in this table. Numbers of men in the Low, Mid, and High Hy groups respectively are 80, 81, and 86 and the numbers of women are 44, 39, and 37 respectively.

Table 19-B. Step Four of regression modeling of Fibroplastic Coronary Intimal Thickenings for Low R_a, Mid R_a, and High R_a groups in pools of *basal* men and women aged 15-79 years.

R_a Range	Step Three	Step Four; constant α					R^2		
	b_3/b_2	β	c_1	c_2	c_3	c_4	Total	Simplified	% Retained
			--- MEN ---						
Low	-0.00974	0.196	*	*	*	*	0.7612	0.7583	99.6
Mid	-0.01079	0.213	*	*	*	*	0.8460	0.8439	99.8
High	-0.01041	0.251	*	*	*	*	0.7906	0.7894	99.8
			--- WOMEN ---						
Low	-0.00562	0.103	*	*	*	*	0.8782	0.8633	98.3
Mid	-0.00789	0.104	*	*	*	*	0.8856	0.8735	98.6
High	-0.00827	0.141	*	*	*	*	0.8426	0.8402	99.7

In each instance, the equation $y = \beta Z + c_1A + c_2A^2 + c_3A^3 + c_4R_a$ was determined by regression, where $Z = A^2(1 + \alpha A)$ and $y = F$. The tabulated values of β are those when only βZ enters the equation. The symbol "*" indicates regression coefficients that are rejected as not of statistical significance on further entry after βZ. Total R^2 refers to the entire equation including all insignificant terms, the simplified R^2 refers to the equation retaining only βZ, and % retained is their ratio expressed as a percentage. In each instance, the average ratios of b_3/b_2 obtained at Step Three of model building are used to estimate the parameter α, as reproduced from Table 19-A.

CHAPTER TWENTY

TWO PATHWAYS TO ATHEROMA VARIABLY LINKED TO RENOVASCULOPATHIES

Abstract. Findings in foregoing chapters can be used to introduce the two schemata, (a) Hy ➔ F_C ➔ YesA ← Hy and (b) R_a ➔ YesA ➔ C_T. This chapter performs tests upon certain predictions that arise from the proposed schemata. Those tests fail to invalidate the predictions, thereby adding credence to the schemata.

The hyaline form of renovasculopathy, Hy, was formerly found to associate with the ominous F_C-linked aspect of coronary intimal fibroplasia, while the fibroplastic renovasculopathy, R_a, was associated with the less threatening C_T-linked aspect. These associations led to predictions about the ways that renovasculopathies ought to correlate with coronary atheroma, especially in a multivariate setting.

20.1. Intrusion of atheroma in relation to renovasculopathies

Each of the renovasculopathies, Hy and R_a, has a correlation with atheroma in the coronary artery. As seen graphically in Figure 20-1, the open symbols representing YesA cases, circles and triangles, tend to have the highest values of Hy or of R_a or of both. The downward sloping solid and dashed lines, depicting men and women respectively, delineate the optimum separation of YesA from NoA cases when the weighted average, W_A, equals zero in Equation 20-1. The near identity of the two lines is visible in the graph, in keeping with the finding that the sex groups do not differ significantly in this relationship, and can be served by the same discriminant function, namely Equation 20-1. (Symbols are detailed in Chapter One, Article 1.7.)

$$W_A = 0.48Hy + 0.50R_a \quad (T^2 = 0.56 \quad \% = 63 \quad c=-1.44) \qquad (20\text{-}1)$$

Cases falling rightward and upward from the sloping lines are predicted to be YesA, and 74 of the 127 YesA men (58 %) are correctly classified in this way. Cases falling below and leftward are predicted to be NoA, and 43 of the 64 NoA men (67 %) are correctly classified in this way. For women, 59 % of YesA and 63 % of NoA subjects are correctly classified.

The weighting of Hy in Equation 20-1 is nearly identical to that of R_a. The two variables add equal and independent influences upon the weighted average (the interaction term Hy$*R_a$ is rejected as of no statistical significance).

20.2. ASCVD in relation to renovasculopathies

Subjects dying of arteriosclerotic cardiovascular diseases, ASCVD, are shown with triangles in Figure 20-1; all are YesA cases. When contrasted with the YesA cases

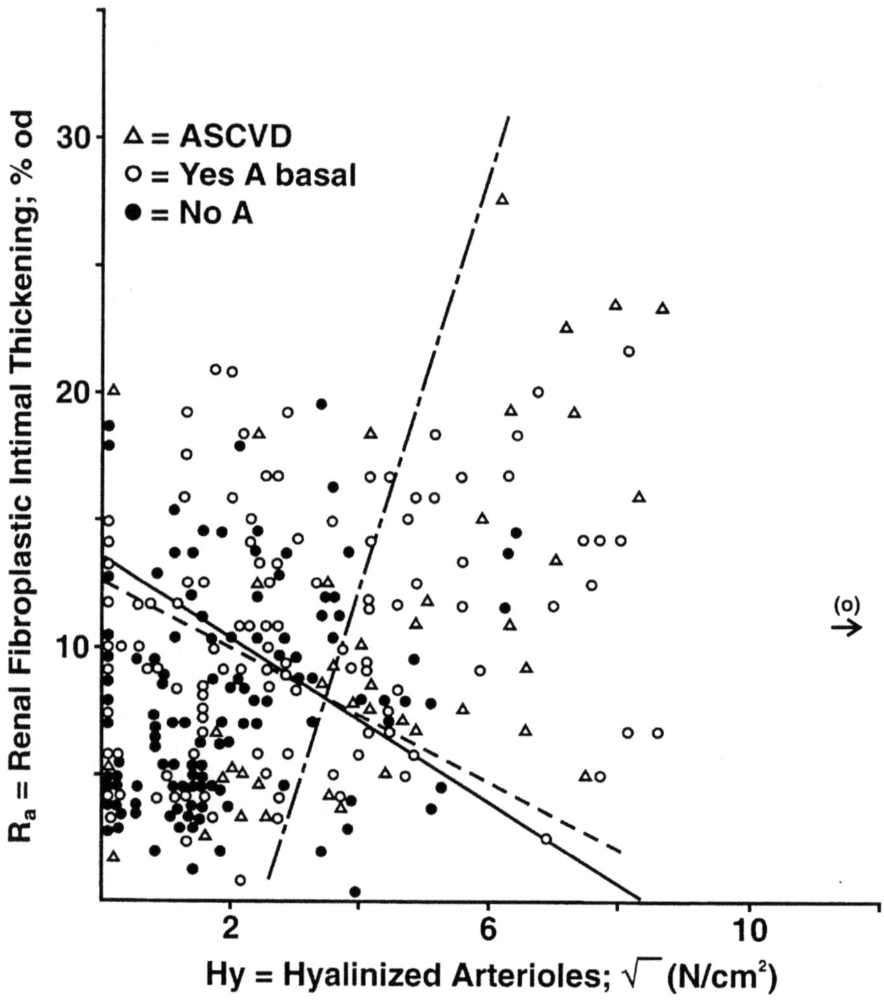

Figure 20-1. Downward sloping lines represent optimal discriminations of open symbols (circles and triangles combined) from dots, separately determined for men (solid) and women (dashed); these are versions of Equation 20-1 with W_A = 0. Nearly vertical broken line optimally discriminates triangles from circles; this represents Equation 20-2 with W_D = 0.

triangles in Figure 20-1; all are YesA cases. When contrasted with the YesA cases without ASCVD (circles), those with ASCVD (triangles) show exceptionally severe hyalinization of renal arterioles (25 of 42 = 60 % of ASCVD subjects fall rightward of the nearly vertical broken line compared with 39 of 114 = 34 % of *basal* YesA subjects). The nearly vertical broken line represents Equation 20-2 which optimally distinguishes YesA subjects with and without ASCVD when $W_D = 0$. Again the same function suffices for men and women alike.

$$W_D = 0.40Hy - 0.11R_a \quad (T^2 = 0.18 \;\; \% = 64 \;\; c=-0.39) \qquad (20\text{-}2)$$

Equation 20-2 allots R_a a statistically insignificant weighting in the average, W_D ($p > 0.1$). Hy receives a significant coefficient, in keeping with the right-to-left separation of triangles from circles visible in the graph. This result attributes importance to Hy but not to R_a at this final stage of pathogenesis when ASCVD supervenes upon a circumstance of established atheroma.

20.3. Fatty streak elements in relation to renovasculopathies

The discussion accompanying Figure 1-2 in Chapter One introduced the measure P_F as a reflection of fatty streak elements visible in H&E stained paraffin sections of coronary artery. The High P_F subjects were defined as those above average in the percentage of the tissue sample exhibiting foam cell infiltrates, while Low P_F subjects are below average in each 10-year age group. Equation 20-3 is the discriminant function for optimally separating the High P_F from the Low P_F men, using a weighted average of the renovasculopathies. The comparable equation for women rejects both coefficients and is therefore omitted here.

$$W_P = 0.41Hy - 0.16R_a \quad (T^2 = 0.16 \;\; \% = 59 \;\; c=-0.21) \qquad (20\text{-}3)$$

20.4. Coronary fibroplasia in tertiles of R_a by tertiles of Hy

Table 20-1 serves to introduce the topic of mulivariate analyses: (Grouping by tertiles within 10-year age groups is explained in Chapter Seventeen, Articles 17.1 & 17.3.) When the men grouped as Low, Mid, and High on R_a are subgrouped into Low, Mid, and High on Hy, they tend to fall on the diagonal of the cross tabulation. The Low-Low group has 36 men and the High-High group has 37 men, while the Low-High has 18 men and the High-Low has 22 men. This observation reiterates the correlation between Hy and R_a that is visible in Figure 20-1 as a progression of symbols upward and rightward from the origin of the graph.

In the marginal averages of Table 20-1, F_C relates significantly to Hy but not to R_a. C_T relates significantly to R_a but not to Hy. Figure 20-2 reproduces these marginal averages in a bar graph to emphasize the conspicuous asymmetries. The pattern for women in Table 20-2 reproduces that for men, with one important exception: coronary fibroplasia per SMC, F_c, fails to correlate with Hy in the women.

Table 20-1. Mean values in *basal* men for measures of coronary intimal fibroplasia and SMC numbers by tertiles of renal arteriolar hyalinization according to tertiles of renal arterial intimal fibroplasia. †

Renal Fibroplasia; R_a group	F; Fibroplasia total µm						
	Hyalinization; Hy group			Unweighted			
	Low Hy	Mid Hy	High Hy	Mean			
Low R_a	160	172	187	173A	ANOVA	F	(.p)
(N)*	(36)	(26)	(18)				
					R_a	3.8	0.03
Mid R_a	174	183	213	190AB	Hy	1.9	0.15
(N)*	(26)	(33)	(22)		Interactions	0.4	0.82
High R_a	189	238	229	219B			
(N)*	(22)	(28)	(37)				
Unweighted Mean	175	198	210				
	A	A	A				

	F_c; Fibroplasia per SMC µm/SMC						
Low R_a	7.1	8.0	8.4	7.8A	ANOVA	F	(p)
					R_a	0.8	0.43
Mid R_a	7.1	7.7	9.2	8.0A	Hy	3.2	0.03
					Interactions	0.5	0.71
High R_a	7.9	8.9	8.7	8.5A			
Unweighted Mean	7.3	8.2	8.8				
	A	AB	B				

	C_T; Total SMC's per sample						
Low R_a	22.9	20.1	22.0	21.7A	ANOVA	F	(p)
					R_a	3.5	0.03
Mid R_a	25.1	23.6	23.4	24.0 AB	Hy	0.1	0.93
					Interactions	0.5	0.71
High R_a	24.6	27.1	25.9	25.9B			
Unweighted Mean	24.2	23.6	23.8				
	A	A	A				

† Marginal unweighted means failing to share an upper case letter (A or B) differ significantly, $p < 0.05$.

* (N) = Number of cases.

Table 20-2. Mean values in *basal* women for measures of coronary intimal fibroplasia and SMC numbers by tertiles of renal arteriolar hyalinization according to tertiles of renal arterial intimal fibroplasia. †

Renal Fibroplasia; R_a group	Hyalinization; Hy group: Low Hy	Mid Hy	High Hy	Unweighted Mean	ANOVA	F	(p)
	F; Fibroplasia total µm						
Low R_a	126	108	180	138A	ANOVA	F	(p)
(N)*	(26)	(9)	(9)				
					R_a	5.0	0.00
Mid R_a	121	148	137	135A	Hy	0.5	0.59
(N)*	(16)	(11)	(12)		Interactions	3.5	0.01
High R_a	215	167	153	178B			
(N)*	(12)	(9)	(16)				
Unweighted Mean	154 A	141 A	157 A				
	F_c; Fibroplasia per SMC µm/SMC						
Low R_a	6.3	6.2	8.5	7.0A	ANOVA	F	(p)
					R_a	0.4	0.65
Mid R_a	6.6	8.0	7.0	7.2A	Hy	0.7	0.44
					Interactions	3.5	0.01
High R_a	8.5	6.8	7.2	7.5A			
Unweighted Mean	7.1 A	7.0 A	7.6 A				
	C_T; Total SMC's per sample						
Low R_a	19.7	18.4	22.0	20.0A	ANOVA	F	(p)
					R_a	3.0	0.05
Mid R_a	19.0	18.5	19.2	18.9A	Hy	0.2	0.85
					Interactions	0.4	0.81
High R_a	24.5	23.4	21.9	23.3B			
Unweighted Mean	21.0 A	20.1 A	21.0 A				

† Marginal unweighted means failing to share an upper case letter (A or B) differ significantly, $p < 0.05$.
* (N) = Number of cases.

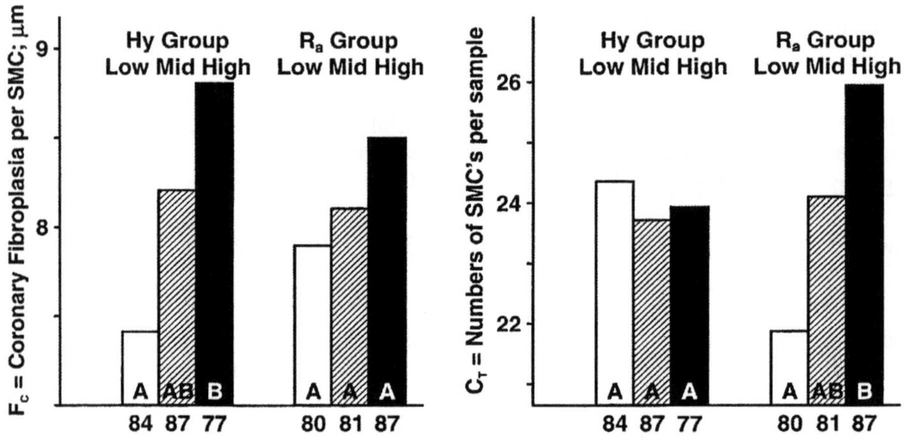

Figure 20-2. These are marginal unweighted means of means from Table 20-1. Columns failing to share a symbol A or B within a triplet differ significantly ($p < 0.05$). Numbers of cases are at bases of columns.

20.5. Multivariate discriminant functions

A four variable system emerged in previous chapters, F_C, C_T, R_a, and Hy. These variables are all found to correlate in assorted ways with atheroma and fatty streak elements in the coronary arteries. It is useful to ask what pathways might the renovasculopathies take to promote atheroma. Two schemata in particular emerge from past chapters and from considerations to be taken up next:

$$Hy \rightarrow ① \rightarrow F_C \rightarrow ② \rightarrow YesA \leftarrow ③ \leftarrow Hy \qquad \text{(Schema a)}$$

$$R_a \rightarrow ④ \rightarrow YesA \rightarrow ⑤ \rightarrow C_T \qquad \text{(Schema b)}$$

Multivariate statistical methods offer opportunities for exploring these proposed schemata. Such methods are hazardous, especially when correlations are weak as in the present circumstances. The hazard arises chiefly from multicolinearity, because placing a third variable between two correlated variables might be only one of many possible arrangements. The following development may have some heuristic value even though much uncertainty attends upon it.

The development followed here is to examine data on men, and later repeat the steps for women. Toward this objective, Equations 4-4 and 20-1 are modified to be reintroduced here for men and women separately. These are discriminant functions drawn to distinguish subjects with YesA from those with NoA coronary arteries. For men we find:

$$W_A = 1.52F_C \quad (T^2 = 1.83 \quad \% = 76 \quad c = -3.25) \tag{20-4a}$$

$$W_A = 0.28C_T \quad (T^2 = 0.08 \quad \% = 53 \quad c = -0.72) \tag{20-5a}$$

$$W_A = 0.38Hy + 0.49R_a \quad (T^2 = 0.39 \quad \% = 60 \quad c = -1.32) \tag{20-6a}$$

20.6. Schema a

An interesting outcome results from introducing F_C into Equation 20-6a.

$$W_A = 0.25Hy + 0.48R_a + 1.49F_C \quad (T^2 = 2.20 \quad \% = 80 \quad c = -4.27) \tag{20-7a}$$

The coefficient for Hy declines, while T^2 rises nearly to match the sum of the two separate T^2 (2.20 vs 2.22 = 1.83 + 0.39 SD^2 units). These results make clear that the major information about atheroma (W_A) resides in F_C, and that the other two variables contribute little in addition. Hy marks a state of accelerated F_C and thereby enhances W_A (Table 20-1). By placing F_C between Hy and W_A the observed shrinkage of Hy's standardized coefficient is expected. However, the coefficient does not vanish, and indeed remains significant. Hence we find reason to suspect an action by Hy directly upon W_A in addition to the indirect action, as depicted in Schema a.

20.7. Schema b

The addition of C_T to Equation 20-6a yields

$$W_A = 0.38Hy + 0.38R_a + 0.23C_T \quad (T^2 = 0.44 \quad \% = 62 \quad c = -1.70) \tag{20-8a}$$

In this setting, T^2 again nearly equals the sum of the two separate sources (0.08 + 0.39 ≈ 0.44). The coefficient of R_a declines, as expected from the finding in Table 20-1 that C_T correlates with R_a and therefore contains some of its information about atheroma. The arrow in Schema b is drawn from YesA to C_T because of information in Chapters Five through Eight. The evidence suggests that C_T might be spuriously elevated in the YesA state through selective censoring of the least cellular locations in an artery. Accordingly, the YesA condition, whether accelerated by Hy or R_a, is expected to elevate C_T, the measure of average SMC numbers in the uncensored locations.

20.8. Contrasting results for women

Repetition of the preceding development with data on women yields:

$$W_A = 3.31F_C \quad (T^2 = 3.97 \quad \% = 83 \quad c = -6.87) \qquad (20\text{-}4b)$$

$$W_A = 0.15C_T \quad (T^2 = 0.02 \quad \% = 55 \quad c = -0.30) \qquad (20\text{-}5b)$$

$$W_A = 0.45Hy + 0.71R_a \quad (T^2 = 0.60 \quad \% = 67 \quad c = -1.77) \qquad (20\text{-}6b)$$

$$W_A = 0.86Hy + 0.00R_a + 3.61F_C \quad (T^2 = 4.90 \quad \% = 88 \quad c = -8.29) \qquad (20\text{-}7b)$$

$$W_A = 0.45Hy + 0.70R_a - 0.04C_T \quad (T^2 = 0.64 \quad \% = 64 \quad c = -1.64) \qquad (20\text{-}8b)$$

The test of Schema b has generally a similar outcome for women as for men, although C_T has an appreciably weaker association with W_A in women than in men. The test of Schema a, however, is sharply different between men and women. Hy fails to lose importance for women as it does for men in Equation 20-7 (instead it is R_a that loses importance, cf. Note below).

The failure of Schema a in women is in keeping with the findings in Table 20-2. F_C in women does not correlate with Hy in women as it does in men. Therefore, the YesA status cannot be displaced from its association with Hy by the addition of F_C to the equation. This finding rejects the placement of F_C into an intermediate position for women, as required in Schema a.

The evidence supports the conclusion that the arrows labeled 2 to 5 in Schemata a and b are equally active in men and women, but that arrow 1 occurs only in men. This deficiency is statistically large enough to fully "explain" the sluggish emergence of F_C and the subsequent laggard progression to atheroma in women.

(Note: In Equation 20-7b, the coefficient of R_a loses significance. This result is paradoxical, because R_a is uncorrelated with F_C in Table 20-2. However, the product moment correlation between R_a and F_C is $r = 0.398$, which is strongly significant. The loss of information due to grouping of R_a into arbitrary intervals, in this setting, happens to obliterate a significant correlation. The existence of that correlation raises the possibility of a third schema, $R_a \rightarrow F_C \rightarrow$ YesA. Perhaps the absence of Schema a in women might allow the emergence of this alternative pathway to atheroma that is overridden in men by the stronger Schema a. Comparison of Tables 19-1 and 20-1 further supports this proposal. In men, F_C correlates significantly with R_a in Table 19-1, but the correlation loses significance in Table 20-1 when Hy is introduced to the multivariate setting, as expected if the stronger Schema a overrides the weaker Schema c. Hence, Schema c emerges only in women, where the arrow 1 of Schema a doesn't apply.)

20.9. Postscript to Tables 20-1&2

Curve fitting was done to relate F to age for men and women within the nine subgroups of Hy by R_a, as done in Chapter Nineteen, Tables 19-A&B. At Step Three, the ratios b_3/b_2 ranged from -0.00864 to -0.01182 in men and from -0.00557 to -0.00953 in women. At Step Four, the expression, $F = \beta Z$ captured all of the statistically significant information of Equation 14-1 in every instance, and retained an average of 98.1 % (93.0 to 99.9 %) of the R^2 of Equation 3.

20.10. Regression modeling of Hy and R_a effects on coronary intima

The development of Chapter Seventeen, Tables 17-6&7, is repeated here with $y = F$, $y = F_C$ and $y = C_T$ in Equation 19-1for $X = Hy$ and then for $X = R_a$. Elimination of insignificant terms yielded for men $F = 0.207A^2 - 0.002334A^3 + 0.00330HyA^2$, $F_C = 0.00383A^2 - 0.0000372A^3 + 0.000130HyA^2$, and $C_T = 0.954A - 0.009424A^2 + 0.438R_a$; for women $F = 3.01A + 0.00100\,R_aA^2$, $F_C = 0.00161A^2$, and $C_T = 0.828A - 0.00712A^2$. These simplified forms of Equation 19-1 retained respectively 98.6, 99.4, 99.4, 98.2, 96.1, and 99.4 % of the total R^2. These results reiterate the correlations of C_T with R_a and of F_C with Hy, but only in men.

20.11. Expanded findings of uncertain significance

Equations 20-9 to 20-11 are constructed for a pool of men and women, and are offered here for general interest, even though complex intercorrelations of these variables leaves the findings without clear interpretations.

$$W_A = 1.93F_C + 0.61C_T + 0.38Hy + 0.34R_a \quad (T^2 = 2.99 \quad \% = 81 \quad c=-6.56) \quad (20\text{-}9)$$

The coefficients of Hy and R_a are diminished substantially, 21 % and 32 % respectively, from their values in Equation 20-1, 0.48 and 0.50. Indeed, the coefficients for Hy and R_a are near to losing statistical significance ($p = 0.03$ and $p = 0.05$ respectively), and they lose 60 % and 68 % respectively of the R^2. These losses are consistent with the schemata a and b.

$$W_D = 0.17F_C + 0.47C_T + 0.41Hy - 0.11R_a \quad (T^2 = 0.40 \quad \% = 65 \quad c = -2.12) \quad (20\text{-}10)$$

The term $0.17F_C$ lacks statistical significance ($p > 0.05$), while the coefficient of Hy loses only 2.5 % in magnitude from Equation 20-2. Hence, the correlation of Hy with ASCVD (Equation 20-2) cannot be "explained" by indirect association with F_C or C_T. The term R_a remains consistently insignificant in both equations. In this instance, the multivariate test serves to invalidate the suggestion that Hy accelerates ASCVD solely by its association with F_C. Between the kidney (Hy) and the heart (ASCVD), something has escaped our attention. A clue to the missing information might be in arrow 3 of Schema a. The observed variable F_C is an average, and could be underweighting some isolated foci of ominous fibroplastic arteriosclerosis of the kind observed in Chapter Eleven.

$$W_A = 2.36F_C + 0.58C_T + 0.39Hy + 0.33R_a + 0.70P_F \quad (T^2 = 3.92 \quad \% = 83 \quad c=-7.89) \tag{20-11}$$

With the addition of the P_F term, the other coefficients are little changed from those in Equation 20-9. This outcome implies that the foam cell effect on atheroma is in large part independent of the other variables in this data set, an implication further supported by the rise of T^2 from 2.99 to 3.92 SD^2 units.

CHAPTER TWENTY ONE

AGE OF ONSET OF THE SEX DIFFERENCE IN CORONARY FIBROPLASIA

Abstract. Intimal fibroplasia in the right coronary artery of men fails to differ significantly from that of women at age 20 years, but comes to exceed it by increasing amounts after age 30 years. Data examined in this chapter raise interesting suggestions: (1) Major events appear to evolve chiefly at premenopausal ages giving no reason to invoke estrogen effects. (2) Coevolution of the two variables P_F and F_C supports the hypothesis that fibroplasia may act as a lipid trap. (3) No support is found for significance of intimal cell masses of infancy.

Intimal and medial fibroplasia of the right coronary artery are generally greater in men than in women throughout most of adult life. The onset of the difference appears to be in the third decade of life, peaking in the fourth or fifth decade, and vanishing or reversing after age 65 years. It would be useful to know the precise patterns of these evolving differences.

21.1. Construction of sliding averages

The cases supplying the specimens for this analysis are described in Chapter One, Article 1.9. Those cases are supplemented with additional cases assembled by ongoing collection, following the same rules. Black and white racial groups show no significant differences in the patterns reported here, and these are pooled into two groups, men and women, to assure maximal numbers of cases in the comparisons to be made. Only *basal* subjects are used, because it is important to have autopsies as nearly as possible representative of the living populations.

Small numbers of cases limit the precision of the determinations that we seek. To overcome this limitation, the technique of sliding averages is used to smooth the otherwise erratic patterns. Means and t-tests are calculated for men and women of ages 15-29 years, then for ages 16-30 years, and so on up to 72-86 years. Results are reported in graphic form. The means of 15-year age groups are not necessarily the same as those for narrower age ranges, of course. But perhaps this rough beginning can illustrate how more definitive data might be assembled.

21.2. Coronary intimal fibroplasia per SMC in men and women by age

The variable F_C is described in prior chapters. It is the fibroplastic collagenous matrix associated on average with each of the intimal SMC's. The means of this variable for men and women are plotted for sliding 15-year age groups in Figure 21-1. No appreciable sex difference is visible until age 30 years, and this reaches statistical significance at age 33 years ($p < 0.05$) or age 35 years ($p < 0.01$). The difference reaches maximal at about age 40 years, and shrinks to insignificant by age 47 years (p

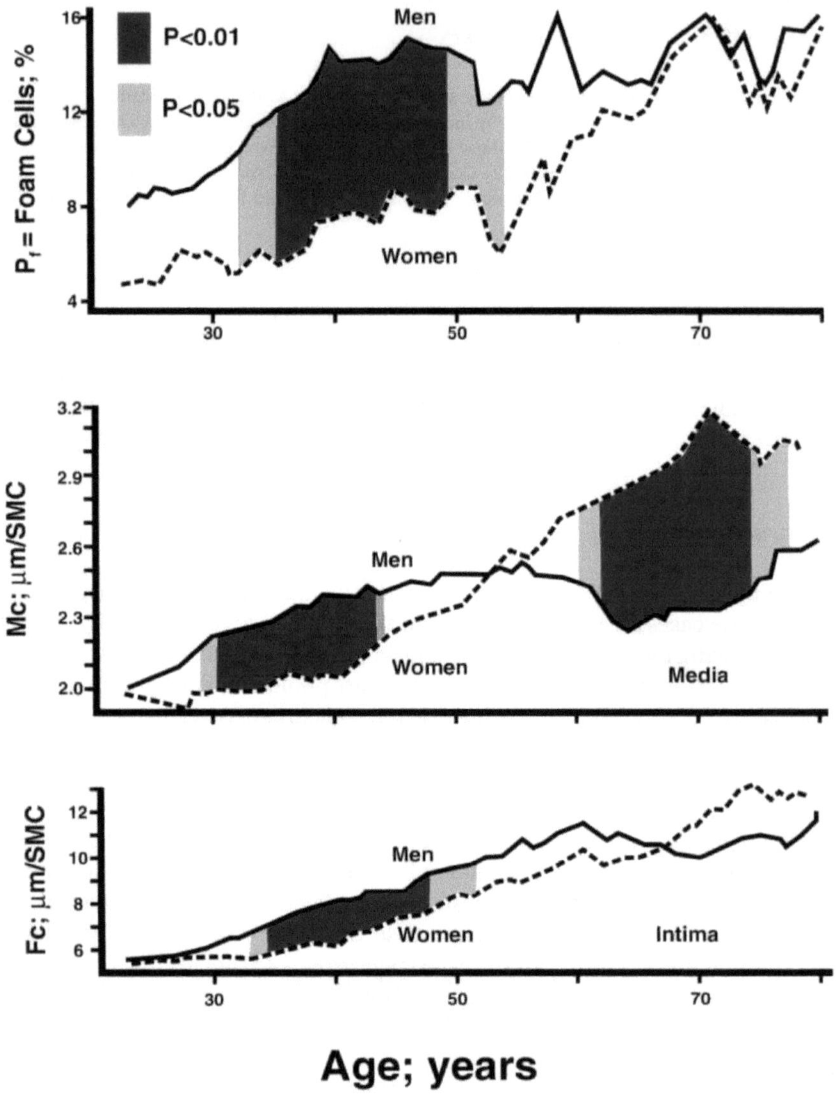

Figure 21-1. Sliding averages of 15-year age groups are reported for basal men and women.

> 0.01) or age 52 years (p > 0.05). There appears to be a reversal after age 65 years, but the result is not statistically significant.

21.3. Coronary medial fibroplasia per SMC in men and women

The variable M_C, medial fibroplasia per SMC, is introduced in Chapter Sixteen. Men come to exceed women in M_C at younger ages than happens for the intima (Figure 21-1). The difference reaches significance by age 28 years ($p < 0.05$), and is visible before that age. The peaking (around age 35 years) and subsequent decline to insignificance (age 44 years) of the sex difference also happen sooner in the media. Furthermore, the late life reversal obtains statistical significance for the media but not the intima. These findings might tend to suggest a process beginning in the media and later extending into the intima. The result is a surprise, however, not anticipated prior to the analysis. It must therefore be held with great caution until confirmed by independent testing.

21.4. Sex difference in F, the total coronary intimal fibroplasia

Total intimal fibroplasia, F measured in micrometers, is the intimal thickness averaged over sites that lack atheroma. The sex difference in this variable, male minus female, is plotted in Figure 21-2. This chart emphasizes the occurrence of peak sex difference around age 36 years. Beginning at that age, women progressively overtake men, to obliterate the significant difference by age 60 years. For 15 or 20 years, from ages 36 to 55 years, women experience the same acceleration of fibroplasia as the men did at younger ages. This acceleration occurs at a time of life when estrogens are fully active, long before the usual onset of menopause. It is therefore difficult to attribute these differing patterns between the sexes to any purported estrogen effect.

21.5. Fatty streak elements in coronary intima of men and women by age

The pattern for fatty streak elements, measured by the extent of foam cell involvement in nonnecrotic intima, P_F, resembles that for fibroplasia per SMC, F_C, Figure 21-1. The male excess over females achieves statistical significance around age 32 years ($p < 0.05$) or 35 years ($p < 0.01$). It peaks around age 40 years, shrinking thereafter to lose significance after age 48 years ($p > 0.01$) or 54 years ($p > 0.05$), then vanishes in old age. The findings relate to those given in Chapter One, Equations 1-1 and 1-2, where F_C was examined as a possible precursor of P_F and of the sex difference in P_F. Those former analyses raised the suggestion that fibroplasia might precede and predispose to foam cell deposition. That hypothesis is now found to be consistent also with the temporal events in the coevolution of the two variables. The increase in P_F with age, and the excess of this increase in men compared with women could easily be explained by the action of fibroplasia as a lipid trap.

21.6. Intimal cell masses in childhood

Reference [10] offers a penetrating review of this topic. Several investigators have

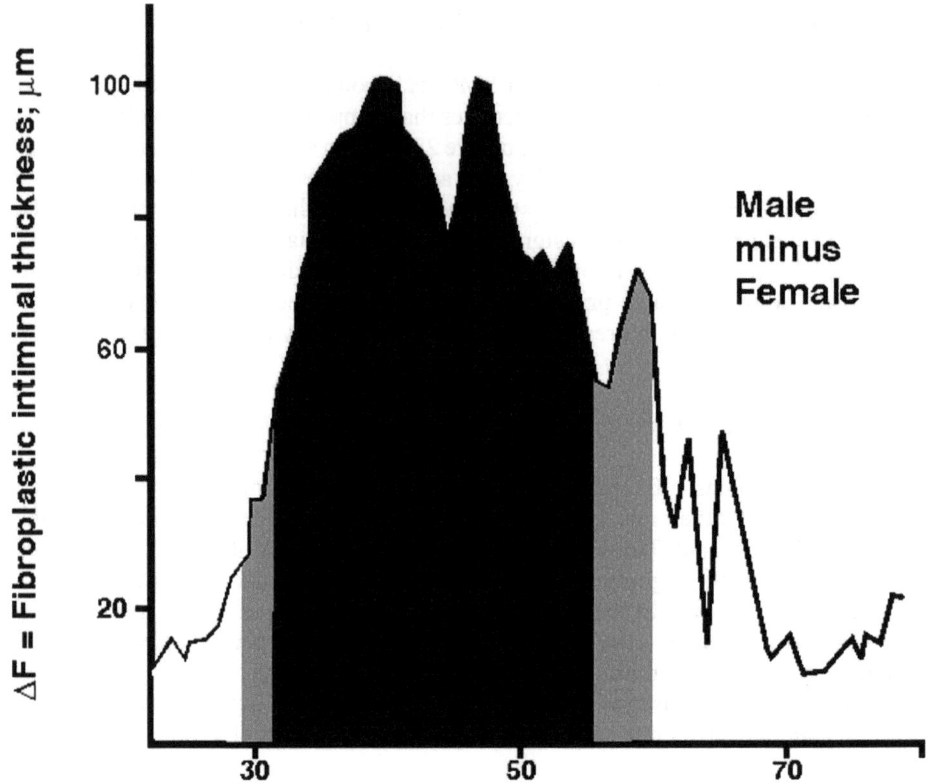

Figure 21-2. Sex difference in F as seen in sliding averages of 15- year age groups. Grey and black areas are as in Figure 22-1.

examined coronary arteries of infants and young children for the presence of newly forming intima, often called "intimal cell masses". Claims are made that the anatomic locations of these masses coincide with the sites of plaque formation in later life. Furthermore, the cell masses were sometimes more abundant and better developed in males than in females. Those observations initiated the persistent opinion that plaque growth begins in infancy with the precocious appearance of intimal organogenesis.

The data in Figures 21-1 & 2 reveal little if any excess of intimal fibroplasia in men compared with women at ages 20 - 25 years. Rather, an accelerated divergence of the sexes seems to begin only after age 25 years. These results do not support the proposal that intimal thickenings, beginning more often in infant boys than girls, persist as an important sex difference into maturity. That outcome, in turn, raises doubts about the significance of infantile cell masses upon later events.

Chapter Twenty Two

Adrenocortical nodularity in relation to coronary fibroplasia

Abstract. The aging adrenal gland undergoes progressive structural changes. These are readily visible at autopsy, although of overly small size to visualize by clinical instrumentation. In the preliminary series of forensic autopsies reported here, the aging features of the adrenal gland were found to correlate with coronary atheroma, and with generalized intimal fibroplasia. The results provisionally place the adrenal cortex into the picture of accelerated aging of arteries, and of their disposition to atheroma.

When the adrenal glands were examined systematically in the series of forensic autopsies reported here, they were found to grow larger from ages 15 to 45 years and then to stabilize or shrink with further aging. Accompanying the enlargement was a progressive cortical replacement by nodules of increasing number and size. Nodularity continued to increase up to age 64 years, and then to stabilize into old age. These features occurred in men at rates equal to those in women. When matched for age, subjects with coronary atheroma averaged greater adrenocortical nodular hyperplasia than those without atheroma. The nodularity also correlated with the aging-associated intimal fibroplasia of the coronary artery. These bits of information originate from the study reproduced here from a recent publication, which is the only source as yet available for data on these topics [76].

22.1. Quantifying adrenocortical nodular hyperplasia

Figure 22-1 defines the method for assigning a quantitative severity grade to the pair of adrenal glands obtained at autopsy. The glands as illustrated in the photograph were formalin fixed, dissected free of fat and connective tissue, weighed and measured, sectioned at 0.5 mm, and the sections attached to plastic sheets with Superglue. The specimens prepared in this way from 96 autopsies were arranged on a table in order of increasing cortical nodularity. The consensus of two observers grouped these specimens into the intervals that were just perceptibly distinguishable, and the photographed examples in Figure 21-1 were chosen to represent their respective grades of severity. Each new specimen is then assigned the number of the image that it most nearly resembles.

22.2. Subjects of the study

Adrenal glands from a series of 96 black and white men and women of ages 16 to 88 years were prepared for assessment as in Figure 22-1. Causes of death were *basal* in 75 cases, coronary heart disease in 15, cerebrovascular in 5, and cardiomyopathy in 1 case. Men of ages 39 to 56 years were of special interest in the findings of the first series, prompting the assembly of a second series of 34 men of these ages (24 of them *basal*). These subjects were obtained as subsets of those used in prior chapters, and were

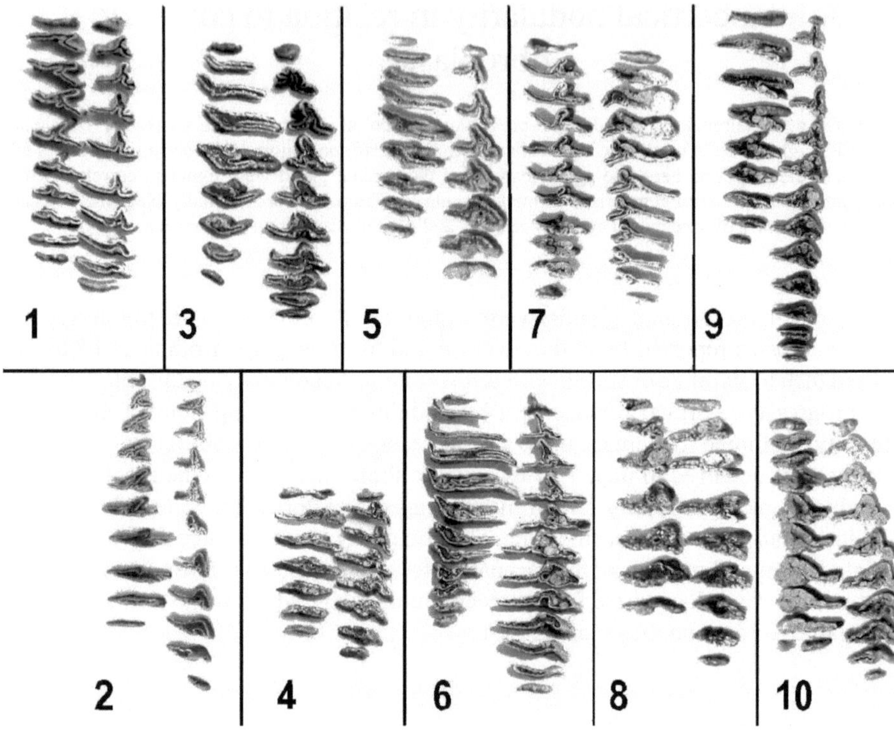

Figure 22-1. Formalin-fixed adrenal gland slices of 0.5 cm thickness stuck to plastic sheets by SuperGlue. Orientation is with patients facing us. Each newly assembled case is assigned the grade of the photograph that it most nearly resembles. Reproduced from reference [76] with permission from Elsevier Science.

evaluated for coronary artery and renovascular features as described in earlier chapters. Details are given in Article 22.13.

22.3. Progression of adrenal features on age

Nodularity of the adrenal cortex increased progressively with age in 10-year age groups from 15 to 64 years, but not above age 64 years, in the data of Table 22-1. Sex and race were rejected for inclusion in the ANOVA model, implying that these factors failed to influence the pattern. The weight of the glands showed no significant increase beyond ages 35-44 years. In these data, nodularity continues to increase into old age while the weight of the glands plateaus or declines. Adrenal gland area calculated as length X width failed to change significantly with age, as reflected in lack of significance of age in the ANOVA for area in Table 22-1. These analyses use the *basal* pool of cases to

Table 22-1. Means by 10-year age group in *basal* Series One and Series Two pool of cases.*†

Age; years	Nodularity; Units	Weight; Grams	Area; Cm^2	Number of cases
15-24	1.6A	3.7A	1.3A	15
25-34	2.3AB	4.2A	1.6B	10
35-44	2.9B	5.4BC	1.5AB	27
45-54	3.3B	5.1BC	1.3A	28
55-64	6.7C	6.2C	1.5AB	9
64-88	6.5C	4.5AB	1.2A	10

* Means within a column that fail to share a letter, A, B, or C differ significantly from each other ($p<0.05$). The ANOVA models reject sex and race for inclusion ($p>0.05$), and model for area rejects age ($p=0.06$).
† Reproduced from reference [76] with permission from Elsevier Science.

approximate the pattern that might be representative of the living population.

22.4. Adrenal nodularity in YesA and NoA subjects; Series One

Subjects with at least one focus of atheroma in the coronary artery sample (YesA subjects) tended to display greater nodular adrenocortical hyperplasia than subjects lacking such a focus (NoA subjects) (Figure 22-2). The association is confounded with age because both nodularity and atheroma increased with age. Prior to age 39 years, instances of nodularity > 3 units were rare, as were instances of atheroma. After age 56 years, both nodularity and atheroma were rarely absent. Only in the age range 39-56 years were all types of cases well represented. In that range, 9 of 17 YesA men showed nodularity > 3 units, while only 1 of 10 NoA men showed such nodularity; the correlation of nodularity with YesA status is significant by Chi Square test ($p<0.01$). Too few women were available for a similar test.

22.5. Adrenal nodularity in YesA and NoA subjects; Series Two

Figure 22-3 presents the association of adrenal nodularity with coronary atheroma in a second series of men aged 39-56 years assembled to retest the issue, along with the Series One data for comparison. In two of three broad age ranges, YesA men significantly exceeded NoA men in adrenal nodularity, when tested by analysis of variance; the excess was not statistically significant at ages < 39 years, represented by

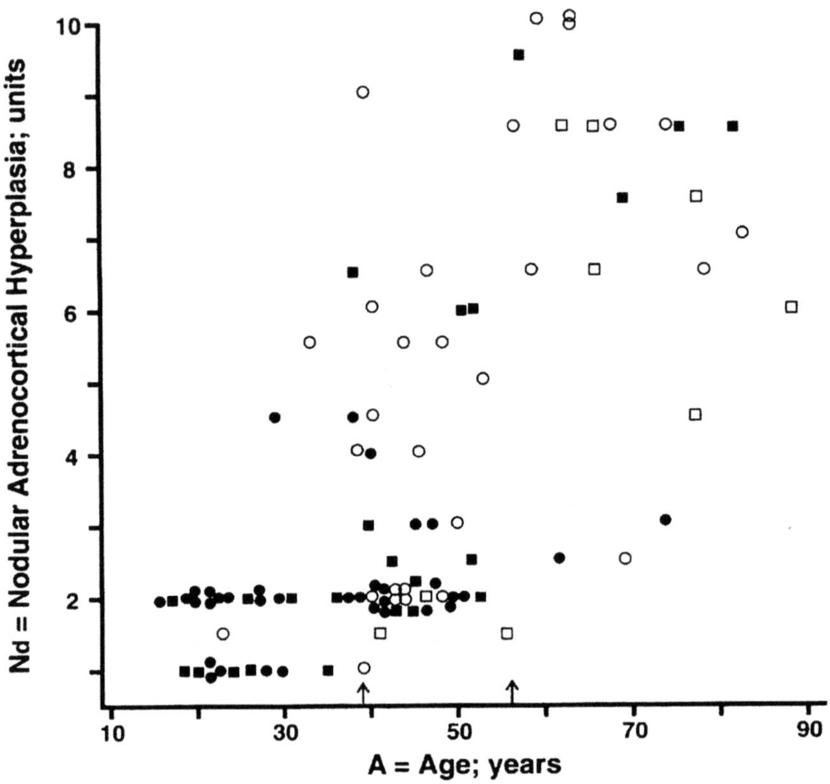

Figure 22-2. Series One (96 cases); average of left and right adrenal glands. Circles are men and squares are women. Open symbols represent YesA cases and filled symbols represent NoA cases. Arrows at ages 39 and 56 years are introduced to aid discussion. Reproduced from reference [76] with permission from Elsevier Science.

only 3 YesA cases. The significant difference obtained at ages 39-56 years in Series One recurred in the replicate test of Series Two. The association of adrenal nodularity with the YesA status was reproducible with acceptable statistical significance even in a series as small as 34 cases, suggesting the possibility of a real biological phenomenon.

22.6. Adrenal weight and area in YesA and NoA subjects

Analysis of adrenal weight and area, as done for nodularity in Figure 22-3, revealed an overall significant difference of YesA from NoA cases in weight but not in area. YesA men averaged 5.68 Gm and NoA men 4.68 Gm ($p = 0.02$) (marginal means in the

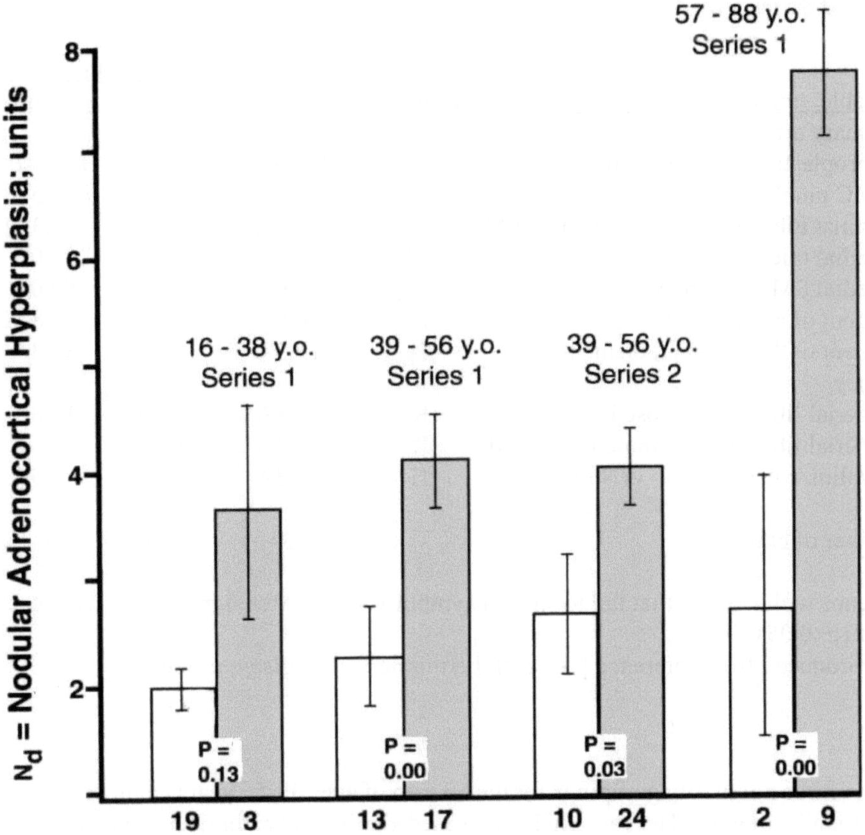

Figure 22-3. Men only. Mean adrenal gland nodularity of YesA subjects (filled bars) compared with NoA subjects (open bars) in three broad age ranges of Series One cases, and one age range of Series Two; y.o. = years old. Vertical line segment atop each bar represents SEM; p levels are significance levels of comparisons between adjacent columns in ANOVA tests.

ANOVA). However, the differences did not achieve statistical significance within any of the individual age ranges as they did for nodularity.

22.7. Coronary intimal fibroplasia and adrenal nodularity

Cases were grouped into Low, Mid and High degrees of nodularity, using cutoff points of $N_d < 3.5$, 3.5-7.0, and > 7.0 units. Table 22-2 reports means of selected variables

Table 22-2. Means of selected variables according to severity of nodular adrenocortical hyperplasia in 60 men age 39-56 years with all causes of death from the pool of Series One and Series Two.*†

Variable; units	Symbol	Nodularity group: Low	Mid	High
Coronary artery				
Fibroplastic intimal thickness; μm	F	233.0A	324.0B	183.AB
SMC numbers; N	C_T	27.8A	28.3A	25.5A
Intimal Fibroplasia per SMC; μm/SMC	F_C	8.5A	11.0B	7.3AB
Medial thickness; μm	M	198.0A	184.0A	206.0A
Medial SMC numbers; N	C_M	83.0A	81.0A	89.0A
Extent of atheroma; %	P_A	10.9A	22.9B	4.8AB
Extent of "fatty streak elements";%	P_F	17.5A	18.3A	30.3A
Kidney				
Arterial fibroplasia close level; %od	R_C	9.8A	10.7A	10.8A
Arterial fibroplasia remote level;%od	R_r	6.2A	8.0A	8.6A
Hyalinized arterioles; $\sqrt{(N/cm^2)}$	H	3.3A	4.3A	3.9A
Number of cases		31	25	4

* Means within a row that fail to share a symbol A or B differ significantly from each other ($p<0.05$).
† Reproduced from reference [76] with permission from Elsevier Science.

within these nodularity groupings for the 60 men of ages 39-56 years having complete data for all variables. The variable F_C increased significantly from 8.5 μm/SMC in the Low nodularity group to 11.0 μm/SMC in the Mid group. The 4 cases in the High nodularity group (F_C = 7.3 μm/SMC) disrupt the pattern of all other cases, and this disruption is taken up next.

22.8. Coronary and adrenal interaction upon atheroma

Men in the age range 39-56 years were pooled from Series One and Series Two, providing a group of 61 cases (only 60 of these having data on all variables listed in Table 22-2). Discriminant function analysis was used to distinguish the YesA from the NoA cases. (Symbols are explained in Chapter One, Article 1.7.) This yielded

$$W = 1.73F_C + 1.07N_d \quad (T^2 = 2.85 \quad \% = 85 \quad c = -5.5) \qquad (22\text{-}1)$$

where N_d is the nodularity score obtained as illustrated in Figure 22-1 and F_C is the

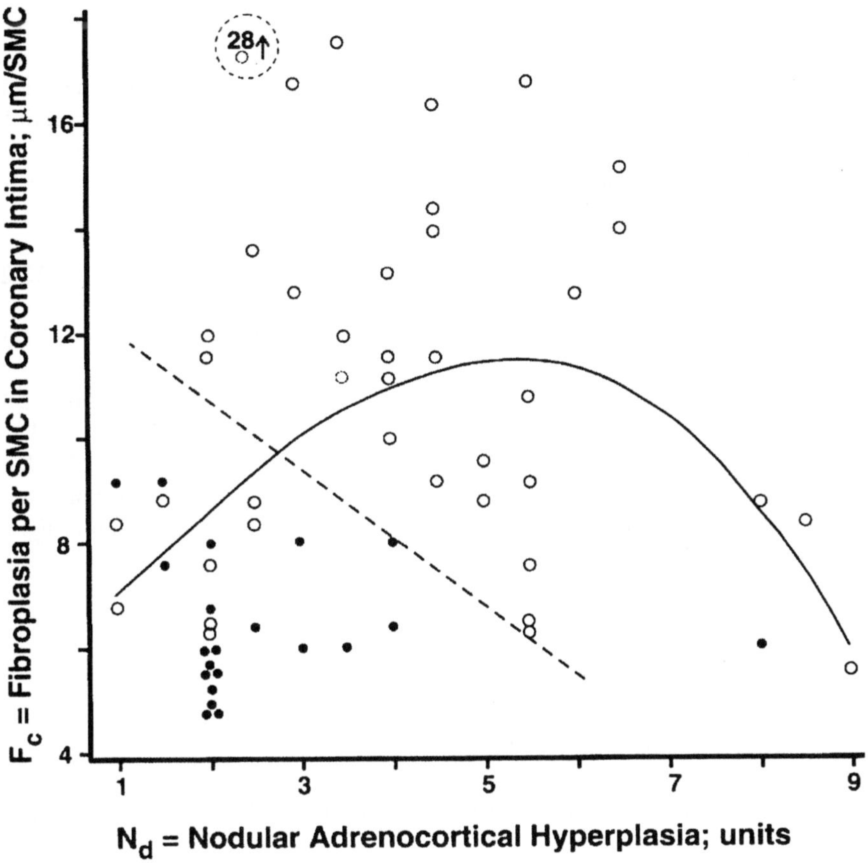

Figure 22-4. Men aged 39-56 years. Open = YesA, filled = NoA. Arching solid curve plots Equation 22-4, sloping dashed line plots Equation 22-1, optimally separating circles from dots. Arrow indicates outlier case with $F_C = 28$ μm/SMC. Reproduced from reference [76] with permission from Elsevier Science.

generalized intimal fibroplasia per SMC as discussed in prior chapters. The standardized coefficients are given; these can be converted to their raw forms, 0.417 and 0.547 by dividing each coefficient by its appropriate standard deviation (this equation is plotted in Figure 22-4 as a downward sloping dashed line). The standardized coefficients assess the relative importance of the two variables, indicating nearly 2:1 greater strength for F_C than for N_d in their relationships to the YesA condition. When $W > 0$ the case is classed as YesA and otherwise as NoA; the classifications done in this way were correct in 85% of the cases.

From Table 22-2 we find that F_C increases on average with increasing adrenal nodularity. This raises the possibility that nodularity in some way reflects an accelerated tendency to intimal fibroplasia. The enhanced F_C is expected to promote atheroma. Hence we derive the hypothetical sequence, $N_d \rightarrow F_C \rightarrow$ YesA. This schema seems plausible because the increment in F_C that accompanies elevated N_d is certainly expected to promote atheroma. The full effect on YesA status associated with N_d is, however, not entirely captured in this schema. The partial failure can be seen by comparing Equations 22-2 &22-3 with 22-1.

$$W = 1.66F_C \quad (T^2 = 1.93 \ \% = 74 \ c = -3.5) \qquad (22\text{-}2)$$

$$W = 0.99N_d \quad (T^2 = 0.83 \ \% = 69 \ c = -1.68) \qquad (22\text{-}3)$$

The sum of T^2 in Equations 22-2 and 22-3 is nearly equal to the T^2 in Equation 22-1, $1.93 + 0.83 = 2.76 \approx 2.85$. Furthermore, the coefficients in Equations 22-2 and 22-3 are nearly identical to those in Equation 22-1. Allowing F_C to enter into Equation 22-3 does not displace N_d from Equation 22-1, implying that N_D is an important determinant of the YesA status beyond just its action mediated through the N_D effect upon F_C. A possible reason for this outcome is taken up next.

22.9. Adrenal nodularity acting upon atheroma by way of F_C

Linear regression was used to determine F_C as a cubic function of N_d. The equation retaining only significant terms ($p < 0.05$) was

$$F_C = 1.72N_d - 0.0203N_d^3 + 5.3 \quad (R^2 = 0.121) \qquad (22\text{-}4)$$

The equation is plotted in Figure 22-4 as an arching curve. The 4 outlier subjects with nodularity grades above 7 units all happen to have low values of F_C, and these outliers cause a downward turn in the otherwise upward sloping curve.

Three of the four cases with $N_d > 7$ were YesA cases, but with low F_C. These 3 influential outliers could excessively contribute to the earlier finding that adrenal nodularity influences atheroma by an effect beyond just its association with F_C. However, after omitting these 4 cases, the coefficients are little changed from those in any of the three equations; F_C still fails to displace N_d from Equation 22-1. The residual influence of nodularity therefore calls for an explanation. A possible explanation of this phenomenon could be that small foci of high fibroplasia might occur within arteries that have overall small average F_C, as observed in Chapter Eleven. It is not necessary for the nodularity effect to fall upon the entire artery, but only upon some small segments of it. That possibility calls for further investigation.

22.10. Renal microvascular features

Interlobular arterial intimal fibroplasia in the renal cortex, measured by R_T and R_c, and the hyalinization of arterioles, measured by Hy, all show no significant correlation with adrenal nodularity (Table 22-2).

22.11. Cause of Death

None of the adrenal gland features differed significantly between cause of death categories (data not shown).

22.12. A speculative synthesis

In company with aging of the coronary intima, resident SMC's tend to elaborate ever increasing amounts of excessive fibroplasia (compare Figure 1A with 1B in Chapter One). The fibroplasia in turn is thought to promote intrusion of atheroma, as elaborated extensively in foregoing chapters. This aging phenomenon, however, is not equally active in all populations or in all subjects within a population. Some unidentified factor or factors must be acting to modify the rates of this aging phenomenon from one subject to another. Perhaps those same factors act upon the adrenal cortex to promote nodular hyperplasia. It is not necessary to imagine an influence mediated through the adrenocortical hormones, although this might be the easiest line of reasoning. The proposal seems of sufficient interest to merit further inquiry.

It seems clear from the data reviewed here that nodularity of the adrenal cortex is in some way, directly or indirectly, linked to the aging changes in the coronary intima. Adrenal nodularity and intimal fibroplasia both increase with age. Those subjects with accelerated nodularity tend also to experience accelerated intimal fibroplasia. This empirical observation is newly obtained, and remains to be explored. A study of adrenocortical nodular hypertrophy might prove to be a study of accelerated aging of the coronary artery. It offers a tantalizing clue to what promising new studies to undertake.

22.13. Selection of cases

22.13.1. *Series One*

Adrenal glands, kidney samples, and coronary artery specimens were obtained from 96 forensic autopsy cases at the Orleans Parish Coroner's Office. These autopsy cases were taken in sequence, omitting only some on busy days when resources were limited. Cause of death was violent in 64 cases, noncardiovascular in 11, coronary heart disease in 15, cerebrovascular disease in 5, and cardiomyopathy in 1 case. Ages ranged from 16 to 88 years, 63 were men and 33 women, 49 were black and 47 were white.

22.13.2. Series Two

Analysis of Series One identified the age range 39 - 56 years as of special interest. To reassess and extend this subsample of cases, a second series of autopsies in this age range was assembled, omitting women because of their small numbers. Cause of death was violent in 16 cases, noncardiovascular in 5, coronary heart disease in 8, cerebrovascular in 2, and cardiomyopathy in 3 cases. Among these 34 men, 18 were black and 16 white.

22.13.3. Basal cases

In Series One, 64 cases of death by violence were combined with 10 cases of noncardiovascular natural causes and 1 of cardiomyopathy. In Series Two, 16 plus 5 plus 3 respectively yields a pool of 24 cases. These two pools of cases were combined to produce a set of 99 cases classed as *basal*, having conditions with no known correlation with atherosclerosis or hypertension. These cases were not caused to enter into the series of autopsies because of arterial diseases, and are therefore approximately representative of the living population [33,34]. The *basal* set of cases is used in the construction of Table 22-1. The properties of the *basal* set are undesirable for the other analyses.

CHAPTER TWENTY THREE

ATHEROMA AND INTIMAL FIBROPLASIA IN PERIODONTAL DISEASE

Abstract. A correlation of coronary heart disease risk with periodontitis is strongly supported by a growing body of clinical studies. The findings reported in this chapter point to the fibroplasia per SMC in the coronary intima as the specific stage of atherogenesis most closely associated with poor dental condition. This finding offers possible clues to the mysterious origins of the fibroplasia.

In 1989, Mattila, et al. [31] noted , " . . . a clinical impression that chronic dental infections are common among patients with acute myocardial infarction . . ." To test the impression they assembled a case-control study having 100 patients with recently diagnosed acute myocardial infarction and 102 matched controls. Those authors constructed a "total dental index" by averaging measures of caries, periodontitis, and absence of teeth; they then applied the index to subjects in the case-control study. They concluded that " . . . dental health was significantly worse in patients with acute myocardial infarction than in controls. The association remained valid after adjustments for social class, smoking, serum lipid concentrations, and the presence of diabetes." Fortunately, the study recruited only subjects less than 60 years of age and mostly men (84%). Without these favorable choices, the association might never have been found.

A large prospective study in 1993 [13, DeStefano et al.] confirmed the claim of Mattila et al., noting that "Dental disease is associated with an increased risk of coronary heart disease, particularly in young men." Using the data base of the National Health and Nutrition Examination Study I (NHANES I) done in 1971-74, over 18,000 subjects were followed for up to 14 years. Dentate subjects were, in one analysis, classed as having or not having periodontitis. Coronary heart disease events were followed by hospital admission and mortality data, and the event rates were adjusted for numerous covariates including smoking, alcohol consumption, obesity, diabetes, blood pressure, and blood lipids. The men and women of ages 25-74 years had overall odds ratio of 1.23 (1.05-1.44, 95% CI) for experiencing a coronary event if periodontitis was present at baseline. Restricting the analysis to men younger than 50 years at baseline (up to 64 years at final followup), the odds ratio was 1.72 (1.10-3.15). The strong effect seen in younger men is clearly influencing the total pool to draw its result upward, meaning that the inclusion of women and older subjects is drawing downward the overall odds ratio. The data imply a strong effect in young men that is absent or greatly diminished in women and older subjects. Mattila et al. in 2000 [32] made a similar observation on their recent negative findings, "We believe that the higher age of the participants in the present study is the most likely reason for the results."

In 2000 [22] Hujoel et al. reported further followup results from the NHANES I study. They followed 8032 eligible subject for an unspecified number of years (perhaps 2000 minus1971 = 29 years in some extreme instances). The authors conclude, "This study did not find convincing evidence of a causal association between periodontal disease and CHD risk." They reach this quizzical decision in spite of the data in their Table 3. The subjects recruited at ages of 35-44 years who were classed as having periodontal disease experienced an odds ratio for new coronary events of 1.80 (1.04 to 3.10, 95 % confidence interval) compared with the disease free subjects having some teeth to inspect. At ages 45-54, 55-64, and 65-74 years respectively the odds ratios were 1.53 (0.94 to 2.47), 1.01 (0.75 to 1.37), and 0.94 (0.68 to 1.29). Odds ratios were adjusted for age, age squared, sex, race, poverty index, marital status, education, systolic blood pressure, diastolic blood pressure, serum cholesterol, diabetes, exercise, body mass, alcohol, smoking and nervous breakdown. The odds ratios 1.80 and 1.53 at ages 35-44 and 45-54 years respectively are nearly identical to the value of 1.72 found for the age range < 50 years by DeStefano et al. The effect is most conspicuous in the 35-44 year-old subjects and diminishes progressively with age thereafter (a phenomenon often seen with many coronary risk factors such as blood pressure and serum lipids). The negative outcome reported by Hujoel et al. arose when subjects of inappropriate older ages were retained in the overall pool, thereby diluting to insignificance the few younger subjects.

A growing consensus was recently expressed by Seymour [44], "In our review article, we concluded as follows: 'there is probably now enough evidence to suggest that there is a relationship between CAD (coronary artery disease) and oral health, and we as dentists should begin to be interested.' Events have moved on since that statement was made and the evidence is accumulating. It remains for causality to be established."

A direct look at the arteries was reported in a recent study [5], where carotid artery intima-media thickness was assessed by B-mode ultrasound. The authors conclude, "The results provide the first indication that periodontitis may play a role in the pathogenesis of atheroma formation, as well as in cardiovascular events." In this chapter, we now extend the inquiry to the level of coronary histology. The findings favor the view that the actions of periodontitis-associated effects may fall directly upon the coronary intima.

23.1. Assessment of dental condition

The autopsy reports on the cases used in Chapter Twenty One sometimes contained information about dental condition. The descriptions of dental condition were done in a consistent way by a single observer throughout the years of collecting specimens. Numbers of teeth were described in terms of how many missing: none, some, or all, with respective values 0, 1 & 2 assigned to them. Periodontal disease and caries were each recorded as absent or present irrespective of severity, with respective values of 0 and 1 assigned to them. These were visual determinations made during anterior inspection of the mouth, often clenched in rigor, and without the use of measuring instruments. The conditions of caries and periodontal disease nearly always

Table 23-1. Descriptive characteristics of selected variables, including correlation coefficients, raw (upper triangle) and age adjusted partial (lower triangle) in 204 men aged 30-59 years.

					Correlation coefficients					
Variable	Symbol	Units	Mean	SD	Nt	Per	Car	Y	F_C	P_F
Age	A	Years	43.8	7.5	.29**	.23**	.21**	.33**	.28**	.12
Nteeth missing	Nt	0,1,2	0.74	0.55	.---	.25**	.29**	.21**	.22**	.21**
Periodontal Disease	Per	0,1	0.47	0.50	.19**	.---	.63**	.17*	.25**	.08
Caries	Car	0,1	0.35	0.48	.24**	.61**	.---	-.01	.27**	.07
YesA	Y	0,1	0.56	0.50	.12	.10	-.09	.---	.56**	.49*
Fibroplasia/SMC	F_C	µm/SMC	9.3	4.0	.16*	.20**	.23**	.52**	.---	.42**
Foam cells	P_F	%	15.6	16.6	.19**	.05	.04	.48**	.41**	.---

* = significantly different from zero at $p < 0.05$, ** at $p < 0.01$

coexisted; only 9 of the 167 men of ages 30-59 years had periodontal disease without caries, and only 1 man had caries without periodontal disease. Therefore, subjects found positive for one disorder were nearly always the same subjects found positive for the other disorder, and correlations cannot be drawn between coronary artery status and each of the dental conditions separately.

23.2. Descriptions of variables

Table 23-1 describes the variables examined in this chapter. The data in that table are drawn from men of ages 30-59 years for reasons to be elaborated later. Notice that the age-adjusted partial correlations are significant between coronary intimal fibroplasia, F_C, and number of teeth missing, Nt, presence or absence of periodontal disease, Per, and presence or absence of caries, Car, while the similar correlations for the YesA condition (presence or absence of atheroma) are not significant.

23.3. Coronary intimal fibroplasia, F_C, in relation to periodontal disease by age

Men having some or all of their teeth present to evaluate were grouped into those with or without periodontal disease. Mean F_C was determined for these groups after stratification into 20 year age ranges, 15-34, 35-54, and 55-74 years. Means were again determined in the 20 year age ranges 25-44 and 45-64 years. This strategy generates means in overlapping 20 year age ranges that accomplish smoothing of the data; such smoothing is needed because numbers of cases are few in narrower age ranges. For purposes of significance testing, the measurements of F_C were age adjusted within each of the 20 year age ranges by regression upon age, age^2, and age^3. In every comparison, subjects with periodontal disease reveal greater F_C than those without periodontal disease (Figure 23-1). The differences are statistically significant in the age range 30 to 59 years, and lack significance outside of that range. For this reason, further analyses, including Table 23-2, are restricted to this age range. The magnitude of the difference is substantial, averaging 32 % greater in the periodontal disease groups. Men without periodontitis must reach age 45 years to attain the level of F_C seen in men with periodontitis at age 30 years. Similar analyses for women reveal no significant differences in F_C between dental condition classes (data to follow), and a chart for women is omitted here.

23.4. Coronary intimal condition, F_C, Y and P_F, in relation to numbers of retained teeth

Subjects without periodontal disease were grouped into the three dental condition classes, those who were missing all, some, or none of their teeth. Subjects with periodontal disease formed a fourth group. The variables F_C, Y and P_F were averaged within these four groups and age adjusted by regression upon age, age^2, and age^3. Women missing all of their teeth had significantly greater F_C than those missing some

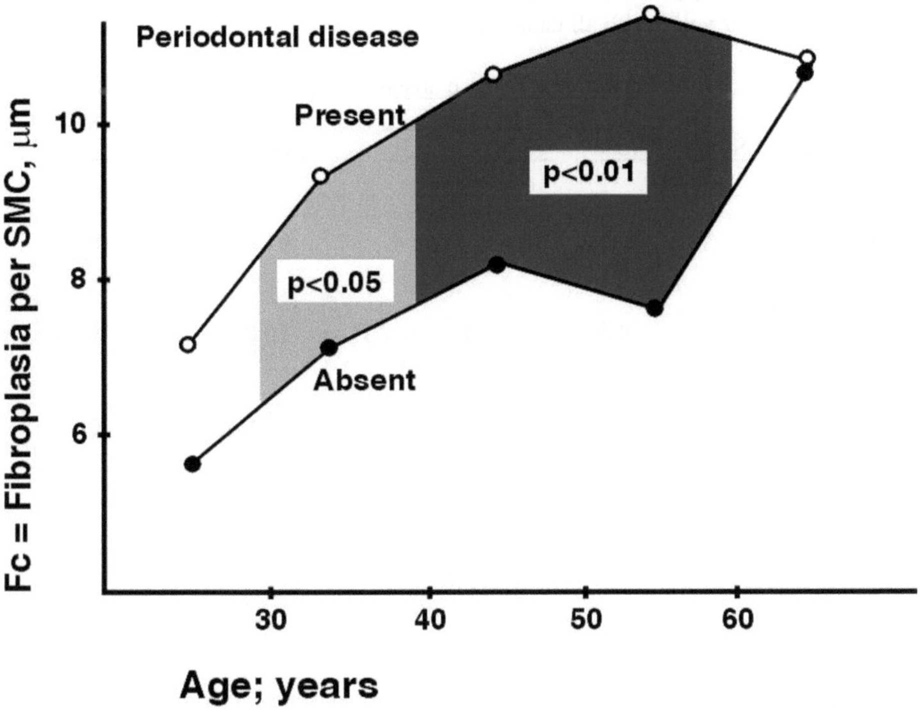

Figure 23-1. Means of overlapping 20-year age groups are used to smooth the data. P-levels of differences between periodontitis classes are determined from ANOVA after adjustment within 20-year age groups for age, age^2, and age^3. Men of all causes of death are included, with women excluded.

or none of their teeth (Table 23-2). The result derives, however, from only 6 edentulous women and with borderline significance in the ANOVA, $p = 0.06$. It would be hazardous to accept this finding as meaningful. Furthermore, the men failed to reveal any such significant findings. The variables Y and P_F also failed to show significant differences in relation to tooth numbers in the absence of periodontal disease. A similar analysis for caries in Table 23-3 gives essentially identical results as for Table 23-2, and this is because the subjects with caries are nearly always the same ones who also have periodontal disease.

23.5. Coronary intimal condition, Y and P_F, in relation to periodontal disease

The occurrence of atheroma, i.e. the YesA condition, is more frequent in men with periodontal disease than in men of the other three dental condition classes. Of the 53 men who retained all of their teeth in a healthy condition (i.e. those noted in Table 23-2

Table 23-2. Means of selected variables in relation to periodontal status in men and women aged 30-59 years with all causes of death.*

Variable	Symbol	Periodontal disease absent Teeth missing None	 Some	 All	Periodontal disease present	ANOVA or Chi Square† P level
			--- MEN ---			
Atheroma 0,1	Y	45.5	56.4	43.4	61.7	p>0.05
Fibroplasia; µm per SMC	F_C	8.0A	9.2AB	8.2AB	10.2B	P<0.01
Foam cells; % of specimen	P_F	11.5A	16.3A	17.5A	16.8A	P=0.28
Number of cases	N	53	45	16	95	
			--- WOMEN ---			
Atheroma 0,1	Y	24.4	19.5	42.3	19.5	p>0.05
Fibroplasia per SMC; µm	F_C	6.8A	7.4A	10.3B	8.5AB	P=0.06
Foam cells; % of specimen	P_F	10.2A	7.5A	6.7A	8.9A	P=0.91
Number of cases	N	42	22	6	18	

* Age adjustment for A, A^2, and A^3, is performed by regression before further significance testing; means within a row failing to share a symbol A or B differ significantly at $p < 0.05$.

† The yes-no variable, Y, is examined by Chi Square test for equal frequencies among the dental health classes; the quantitative variables, P_F and F_C, are examined by ANOVA.

as missing none of their teeth), 45.5 % had atheroma in the coronary artery specimens (age adjusted mean YesA). This compares with 61.7 % of the 95 men who had periodontal disease. The difference is significant by Chi-square test, $p < 0.05$. However, the subjects with periodontal disease did not differ from the other two dental

Table 23-3. Means of selected variables in relation to caries status in men and women aged 30-59 years with all causes of death.*

Variable	Symbol	Caries absent: Teeth missing: None	Some	All	Caries present	ANOVA or Chi Square† P level
		--- MEN ---				
Atheroma 0,1	Y	48.3	70.3	42.4	50.3	P>0.05
Fibroplasia; µm per SMC	F_C	8.0A	9.3AB	8.2A	10.5B	P<0.01
Foam cells; % of specimen	P_F	11.4A	17.2A	17.5A	16.8A	P=0.18
Number of cases	N	61	62	16	70	
		--- WOMEN ---				
Atheroma 0,1	Y	25.3	20.8	42.1	17.4	P>0.05
Fibroplasia; µm per SMC	F_C	6.8A	7.7AB	10.2B	8.3AB	P=0.08
Foam cells; % of specimen	P_F	9.1A	7.7A	6.7A	12.6A	P=0.80
Number of cases	N	41	29	6	12	

* Age adjustment for A, A^2, and A^3, is performed by regression before further significance testing. Means within a row failing to share a symbol A or B differ significantly at $p < 0.05$.

† The yes-no variable, Y, is examined by Chi Square test for equal frequencies among the dental health classes; the quantitative variables, P_F and F_C, are examined by ANOVA.

condition classes, those missing some or all of their teeth. The meaning of this outcome is ambiguous, because statistical significance is borderline and inconsistent for the comparisons of the periodontal disease group with each of the other three classes. A possible association of periodontal disease with the YesA condition is suggested, but not strongly supported. Furthermore, no such result is suggested in the data for women.

The variable P_F clearly shows no hint of association with periodontal disease in either men or women.

23.6. A speculative synthesis

The relationships of periodontal disease to coronary intimal conditions in this data set showed strong, consistent, and statistically significant correlations with fibroplasia in young and middle-aged men, but not in elderly men or women of any age. It is as if the periodontitis-associated effect is an exaggeration of the "maleness" risk factor, enhancing but not mimicking the action of maleness. The association with atheromas is much weaker and less consistent than with fibroplasia. Furthermore, the measure of lipid deposition, P_F, fails to show any persuasive correlation with any measure of dental condition. If these provisional findings can be verified by further studies, then some interesting conclusions would emerge: The stage of pathogenesis that associates best with poor dental condition appears to be fibroplasia, the early, rate-limiting stage, rather than lipid deposition, which usually follows after fibroplasia at a later stage of evolution. If valid, this result could greatly simplify the search for mechanisms that might link dental condition to coronary atherosclerosis. The search could turn attention away from putative actions upon blood lipids or upon established lesions such as fatty streaks or atheroma. This in turn could offer a valuable clue that might open some possible avenues to understanding the mysterious origins of the fibroplasia itself.

CHAPTER TWENTY FOUR

ATHEROMA AND INTIMAL FIBROPLASIA IN RELATION TO OBESITY

Abstract. Measures of obesity in this study revealed associations with coronary intimal lipid deposition, as reflected in the extent of foam cell infiltration seen in H&E stained paraffin sections. The aging linked fibroplasia had no clear or consistent association with obesity. Whereas the fibroplasia of aging seems to be accelerated by factors such as periodontal disease and hyalinization of renal arterioles, lipid deposition seems to be affected by other factors such as obesity. These two dimensions of atherogenesis sometimes act independently of each other in this data set.

In 1983, Solberg & Strong [46] reviewed the information then known about the relationships of coronary risk factors to atherosclerosis determined at autopsy. They noted that, "In most studies no consistent association was found between obesity or physical activity and the degrees of atherosclerosis." In1995 [36], and again in 2002 [35], McGill and associates obtained positive correlations by greatly enlarging the study and by confining attention to victims of violence at ages 15-34 years of age. Those affirmative findings were obtained by measuring atherosclerosis in terms of fatty streaks and raised lesions assessed by inspection of sudan-stained intimal surfaces of formalin fixed arteries. The main findings of relevance to us here are in their Table 4, as reproduced here in Table 24-1. Both measures of atherosclerosis, fatty streaks and raised lesions in the coronary artery, correlated with the obesity measure BMI (body mass index in weight over height squared, kg/m^2) in men but not in women. Findings from the Bogalusa Heart Study [6] subsequently verified these correlations for both fatty streaks and raised lesions in children with prospectively measured risk factor data followed to autopsy, although a comparison of sexes was not reported. A study [27] of coronary arteries obtained from forensic autopsies in 30 women aged 19-40 years confirmed the lack of correlation between BMI and measures of coronary atherosclerosis in these young women, but did find a significant association of abdominal and intraperitoneal fat deposits with atherosclerosis.

Angiography of coronary arteries or ultrasonography of carotid and femoral arteries have repeatedly revealed associations of obesity indices with measures of atherosclerosis [28]. One of those studies [50] examined carotid arteries by ultrasound in 120 men and 187 women of ages 40-70 years, allowing comparison of the sexes. That study reported significant correlation of intima-media thickness with BMI in the men but not in the women. The contrasting findings for men and women seen in Table 24-1 are consistent with these results from clinical imaging studies, but the matter has not been a focus of attention and remains unresolved. In this chapter, we now extend these studies to the level of coronary histology.

24.1. Assessment of obesity

The autopsy reports on the cases used in Chapter Twenty One sometimes contained information about obesity status. A single observer recorded the observations by noting

Table 24-1 Percent intimal area of the right coronary artery involved with lesions by sex and Body Mass Index, BMI, adjusted for race and age*.

BMI kg/m^2	Number of cases	Fatty Streaks % surface	Raised Lesions %surface
	------- MEN -------		
<25	677	3.64	0.96
25-30	324	5.58	1.15
>30	113	9.86	2.44
	------- WOMEN -------		
<25	215	3.81	0.71
25-30	77	6.24	0.82
>30	49	4.60	0.76
ANOVA		p-level	
	BMI	<0.001	<0.004
	Sex by BMI	<0.008	<0.015

* Taken from McGill et. al. Ref. [36]; copyright of this table is owned by the American Heart Association.

the degree of body fatness in a consistent way over the years of collecting specimens. Code numbers were assigned as follows: 0 = wasted, emaciated, or cachectic; 1 = underweight; 2 = within the average range, often appearing optimal; 3 = overweight or excessively fat; 4 = extreme or morbid obesity, ordinarily exceeding average by 50 kilograms or more. These assignments were made by subjective judgement of the amounts of subcutaneous fat, including face and extremities as well as trunk and hips, giving a single judgement to the overall view of the entire body. Recording of height and weight began later than the obesity score assignments and are available for only a subset of subjects. These were used to calculate body mass index, BMI = kg/m^2. Discriminant function analysis was used to find cutoff points for BMI to group the subjects into convenient intervals. Four equations of the form $W = b(BMI) + a$ were obtained to optimally separate obesity class 1 from 0, 2 from 1, 3 from 2, and 4 from 3. Cutoff points for men, obtained when $W = 0$ in these equations, were 20.8, 25.5, 31.1, and 39.8 kg/m^2 respectively, so that BMI class = 0 if BMI < 20.8, 1 if 20.8 < BMI < 25.5, 2 if 25.5 < BMI < 31.1, 3 if 31.1 < BMI < 39.8, and 4 if BMI > 39.8. For women, the respective cutoff points were 17.7, 22.4, 27.8, and 39.0 kg/m^2, the differences from men chiefly reflecting the greater contribution to BMI by muscle mass in the men. The BMI class agreed with the obesity score in 142 of the 236 (60 %); it differed by one unit in 93 subjects (39 %), by 2 units in 2 subjects

Table 24-2. Numbers of cases by age and obesity score and mean obesity score according to age; men and women separately, all causes of death.*

Age; years	Obesity score; 0 to 4: 0	1	2	3	4	Obesity Mean
			--- MEN ---			
15-24	1	33	13	4		1.39ABC
25-34	0	35	19	6	2	1.60BCD
35-44	3	40	46	13	6	1.80D
45-54	5	31	33	9	3	1.67CD
55-64	5	19	13	2	0	1.31AB
65+	9	12	12	2	0	1.20A
			------ WOMEN ------			
15-24	0	9	8	7	1	2.00AB
25-34	1	8	11	15	1	2.19AB
35-44	3	5	16	17	3	2.28AB
45-54	2	4	11	15	5	2.47B
55-64	1	4	5	9	2	2.33AB
65+	2	10	4	7	2	1.88A

* For men and women separately, mean obesity scores that fail to share a symbol A, B, or C differ significantly between age groups at $p < 0.05$. ANOVA for age effect on mean obesity score, men: $F = 4.6$, $P < 0.00$; women: $F = 1.4$, $P < 0.24$.

(1 %), and by 3 or 4 units in no subjects. These disagreements are shown in a later table.

24.2. Relationship of obesity score to age

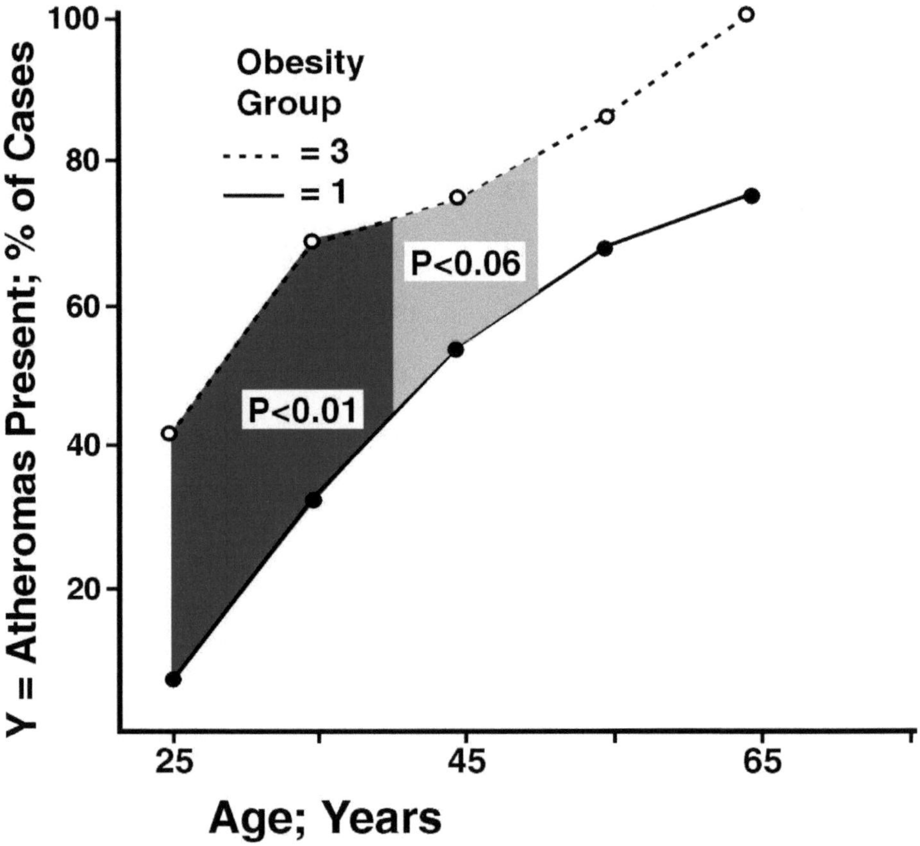

Figure 24-1. Men with obesity scores of 3 (overweight) are compared with those of score 1 (underweight) by ANOVA, means adjusted for age within overlapping 20-year age groups.

In this series of forensic autopsies, obesity scores rise with age from 20 years to a peak in men at ages 35-44 years and in women at ages 45-54 years, falling thereafter into old age (Table 24-2). The trends are small and of weak statistical significance even with substantial numbers of cases (367 men and 188 women). Nevertheless, age adjustments are used in subsequent analyses to minimize any influences that these trends might introduce.

24.3. Influence of obesity on age trends in atheroma

Men with obesity scores of 3 (overweight) are compared with those with scores of 1

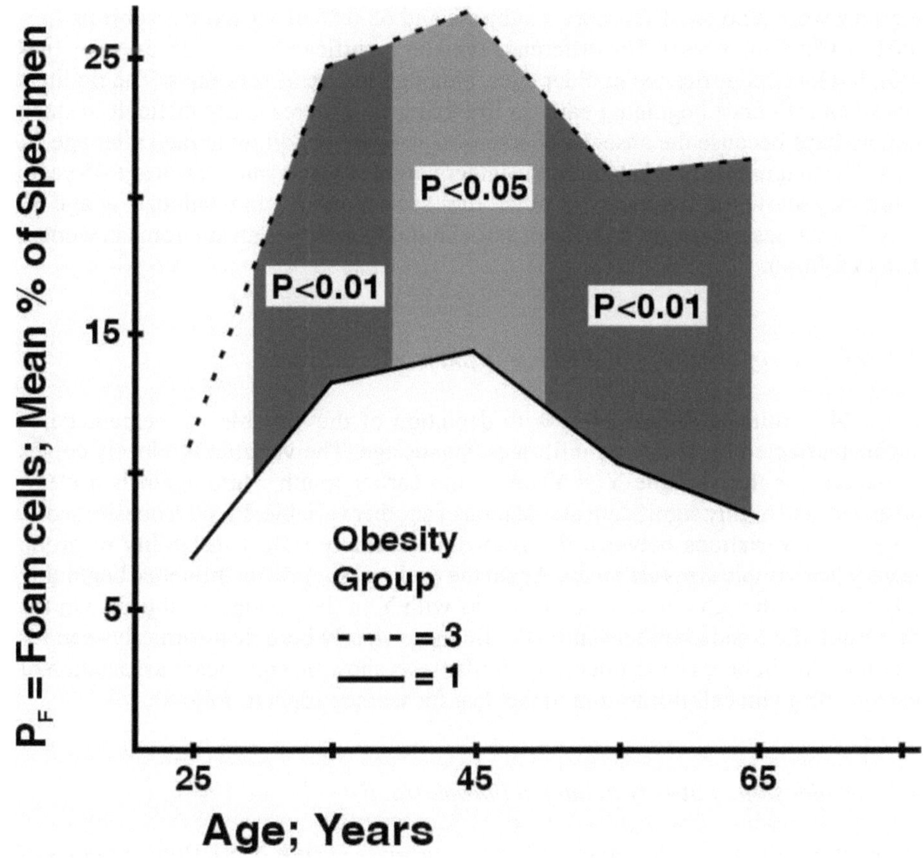

Figure 24-2. As in Figure 24-1.

(underweight) in Figure 24-1. Plotted there are the 20-year sliding averages of the measure YesA (1 = yes an atheroma is present in the specimen, 0 = no an atheroma is not found). Smoothing the data with sliding averages is adopted because numbers of cases are often small in the subgroups plotted in the graphs. The means within 20-year age groups are adjusted to Age, Age^2, and Age^3 in the ANOVA when calculating significance levels for differences between obesity groupings. The graph shows a clear, consistent, and highly significant association of atheroma frequency with obesity. The effect is already present in the youngest age range, 15- 34 years (plotted at abscissa A = 25), where score 3 subjects revealed atheroma in 42 % of 10 cases, while score 1 subjects had atheroma in only 6 % of 68 cases (although Chi Square test is inappropriate because of low values for expected frequencies, these age-adjusted percentages are found to be significantly different by ANOVA). The difference is unequivocally significant at ages 25-44 years where the respective frequencies of atheroma were 32.6 % of 78 score 1 subjects and 68.0 % of 19 score 3 subjects ($p < 0.001$ by Chi Square test). The difference remains significant to ages 35-54 years ($p < 0.06$), but loses significance in older ages, although the trend continues. The findings suggest an influence beginning early in life that grows increasingly difficult to show later, perhaps because the absence of atheroma is a rare condition in men after age 54 years. The magnitude is substantial: the underweight subjects must reach age 45 years before they show the frequency of atheromas seen in overweight subjects at age 30 years. Similar tests show no significant association of obesity with atheroma in women (data to follow).

24.4. Influence of obesity on age trends in foam cell infiltrates

Figure 24-2 imitates Figure 24-1, with depiction of the variable P_F, percentage of specimen affected by foam cell infiltrates, versus age. The variable P_F closely copies the pattern for the variable Y = YesA in the earlier graph. Here again is a clear, consistent, and highly significant association of another variable, P_F, with obesity score. Some minor variations between the graphs may simply reflect instability of group means when sample sizes are small. Again the findings suggest an influence beginning early in life, although not seen so clearly as with Y in the youngest subjects. On the other hand, the trend continues into late life more clearly here than formerly seen for atheroma. Again here as for atheroma, similar tests show no significant association of obesity with foam cell infiltration in the data for women (data to follow).

24.5. Relationship of obesity to intimal fibroplasia, F_C

This relationship is complex, and holds some surprises (Table 24-3). Both extremes of obesity, scores 0 and 4, reveal excessive F_C compared with scores 1, 2, or 3. The 169 score 1 men with mean $F_C = 8.4$ μm differ significantly from the 136 score 2 men with $F_C = 9.5$ μm ($p = 0.018$), but not from the 36 score 3 men with $F_C = 9.5$ μm ($p = 0.22$). The inconsistent difference is small and is not significant within any of the 20-year age group comparisons. If an effect is acting upon F_C between obesity scores 1 and 2 or between scores 1 and 3, then the effect appears to be small and difficult to demonstrate. The paradoxical excess of F_C in the obesity score 0 group is a surprise; it is significant and substantial, and it is not due to a few outliers because the trend is seen in most of the 23 subjects in this group. Furthermore, the score 0 group also shows some excess of foam cells, P_F, compared with groups 1 and 2; although not significant, the effect serves to emphasize the irregularity of the 0 group in the overall pattern in the data.

24.6. Obesity fails to correlate with Y, P_F, or F_C in women

Table 24-3 Means of selected variables by obesity score, all ages, all causes of death, age adjusted by regression to A, A^2, and A^3.*

Obesity Score	Fibroplasia µm/SMC	Atheroma % of cases	Foam cells % of specimen	Number of cases
		--- MEN ---		
0	11.4C	53.2AB	13.3AB	23
1	8.4A	42.3A	10.6A	169
2	9.5B	47.4A	15.1B	136
3	9.5ABC	69.0B	21.7C	36
4	11.1BC	62.5AB	13.0AB	11
		--- WOMEN ---		
0	7.0A	22.9A	7.6A	9
1	8.0A	31.0A	5.1A	40
2	8.1A	28.3A	7.6A	55
3	8.2A	26.1A	10.2A	70
4	8.2A	26.5A	9.3A	14

* For men and women separately, means within a column that fail to share a symbol A, B, or C differ significantly between obesity groups at $p < 0.05$.

Age-adjusted means in Table 24-3 are calculated in the pool of subjects of all ages. This seems appropriate in view of the findings in Figures 24-1 & 2, where trends between obesity groupings are generally similar at all ages. In these means, women show no significant correlation of obesity score with any of the three measures of coronary intimal condition.

24.7. Agreement of BMI with obesity score

Mean values of P_F marked in bold type in Table 24-4 pertain to the subjects with perfect agreement between the obesity scores and the BMI groupings. These subjects can be viewed as the ones with perfect agreements in the duplicate measures of body fat. Departures from these bold-faced diagonals of Table 24-4 are therefore in part reflecting other determinants of body mass, especially muscle mass. Hence, departure leftward within rows indicates less than average muscle mass and departure rightward indicates greater than average muscle mass. Leftward or rightward departures had no significant influences on any of the three variables, Y, P_F, or F_C. This was shown by grouping departures into class 1 if leftward, class 3 if rightward, and class 2 if on the diagonal, and then performing ANOVA on the means between classes. Outcomes all failed to reveal significant differences. These findings support the suggestion that

Table 24-4. Mean foam cell infiltration, P_F (numbers of cases in parentheses), measured in percent of specimen affected, by obesity grade according to Body Mass Index (BMI) group; all ages and causes of death; age adjusted by regression to A, A^2, and A^3.*†

Obesity Score	BMI grade 0	1	2	3	4
			--- MEN ---		
0	**6.9** **(8)**	21.6 (3)			
1	7.9 (9)	**11.8** **(40)**	4.7 (12)	42.4 (1)	
2	10.0 (1)	10.5 (17)	**16.8** **(31)**	12.2 (13)	
3			13.6 (1)	**24.0** **(8)**	
4				11.9 (2)	**30.1** **(2)**
			--- WOMEN ---		
0	**9.2** **(2)**				
1		**2.8** **(13)**	10.8 (5)		
2		7.3 (11)	**12.1** **(14)**	0.5 (3)	
3			6.5 (10)	**8.7** **(18)**	18.4 (3)
4				8.5 (2)	**7.3** **(5)**

* Cutoff points for grouping of BMI were 20.8, 25.4, 31.0 & 39.8 kg/m^2 for men and 17.7, 22.4, 27.8 and 39.9 for women. Bold type indicates the data with BMI identical to obesity score, which account for 60 % of the 236 subjects tabulated here.
† BMI data were available on only some of the cases with obesity score data.

Table 24-5. Mean heart weight for men and women by obesity scores; all causes of death, all ages, age adjusted by regression for Age, Age^2, and Age^3.

Obesity Score	Heart weight; Gm Men	Heart weight; Gm Women	Difference M - W	Number of cases M	Number of cases W
0	306	228	78	24	8
1	371	287	84	171	40
2	420	322	98	140	54
3	520	350	170	39	71
4	578	494	84	11	12

ANOVA	F	(p)
Sex	158.3	0.00
Obesity	31.7	0.00
Interact.	4.0	0.00

muscle mass is of no consequence to these features of coronary artery intima.

24.8. Influence of obesity on renovasculopathies

The measures R_a and Hy, fibroplasia of interlobular arteries and hyalinization of renal arterioles respectively, are explained in Chapter Seventeen. In this chapter, these two variables were found to hold no significant association with obesity scores, either for men or women (data not shown). It was found in Chapter Seventeen that these two forms of renal microvasculopathy held significant linkages to coronary intimal fibroplasia. The new findings are that renal microvasculopathies now share another property with coronary fibroplasia, namely that both sets of variables lack significant correlations with obesity.

24.9. The paradox of sex differences in obesity effects

The data in Table 24-3 indicate a conspicuous obesity effect on atheroma and foam cells in the coronary intima of men but not of women. Data from prospective studies consistently reveal an increased risk of mortality [29] and for coronary heart disease events [21] equally among the obese men and women. Data from the Framingham study, for instance, show that the occurrence of coronary events over a 28 year period of followup was 46.6 % higher in the heaviest 1/5 of men compared the lightest 1/5, and the comparable figure for women was 43.6 %. On the face of it, the findings seem to indicate that the higher risk in men may be traced to an effect of obesity upon atheroma formation in the coronary arteries, while the higher risk in women must be attributed to a different effect, perhaps acting through cardiac hypertrophy or heart failure because of the heavier work load upon the heart of overweight women. The data of Table 24-5 offer some insight into this possibility. The heart weight of women tends to average about 80 grams lighter than in men. Increasing heart weight associates with increasing obesity scores in men to a greater extent than in women, as shown by the significant interaction term of the ANOVA in the table. With obesity score = 1 (underweight but not emaciated), heart weight of men exceeds that of women by 84 grams, while at obesity score = 3 (overweight but not morbidly obese) the excess is 170 grams. The implication is that women have difficulty keeping up with men in the process of cardiac hypertrophy. This laggard condition may indicate a lesser degree of reserve in the female heart which could possibly generate an unfavorable risk for coronary events and mortality.

24.10. A speculative synthesis

The effects of obesity score on coronary intimal condition arose early in life within this data set. The trends were already established in the youngest observed subjects, ages 15-34 years, and became strongly significant by ages 25-44 years. This early onset of the effects is consistent with the coupling to lipid deposition rather than to the age-linked fibroplasia. With periodontal disease (Chapter Twenty Three) the effect required a lag time of about 30 years for sufficient aging to introduce accelerated fibroplasia, thereby finally promoting atheromas in middle age. No such lag was seen for the obesity related effects because they did not rely upon the fibroplasia of aging. In terms of the photographs on the cover of this book, it appears that periodontal disease accelerates the vertical progression within columns while obesity acts upon the horizontal progression within rows through deposition of lipid.

CHAPTER TWENTY FIVE

PAUCITY OF LITERATURE RELEVANT TO SMC NUMBERS AND THE AGING RISK FACTOR

The data reviewed in this book are almost entirely those from personal experiences of the author. Readers may be puzzled by the paucity of reference to data from other laboratories. This puzzle seems to call for some brief comment.

Chazov, et.al. (1986 [9]) in a review of SMC's in human fibrous plaques noted that, "The cellular aspects of human atherosclerosis remain the least understood . . . At present, cell culture is the only way to study cellular manifestations of human atherosclerosis . . . Relatively little is known about the cellular dynamics in healthy and atherosclerotic human arteries compared to the changes in the vessel wall of other mammalian species." Recent reports seldom make this obvious comment any more, but it can be encountered. Schwartz and Murry (1998 [43]), for instance, state, "We know little about the properties of intimal cells or about their origins, despite the extensive evidence that these unique cells provide the site of origin of atherosclerotic plaques." Virmani et.al. (2000 [79]) also observe, "Unfortunately, there are very few articles on the evolution of early intimal cell masses in humans, and none of these clarify their precise pathological mechanism of development."

Articles on how aging acts upon the artery wall to prepare for atheroma can be found mostly in the early decades of the 20th century, and since the 1970's only in the work of Velican and Velican [77,78]. Rarely does a recent article contain even an incidental mention of this matter. For example, only one of the three articles quoted in the previous paragraph contains the word "age", and that only once, in the title of a table giving data on subjects over or under age 50 years.

A likely reason for the widespread neglect of these matters can be found in the writings of many authors, exemplified by those of Stary (2000 [49]): "In the original classification, minimal changes (type II lesions) are subdivided into those at highly susceptible sites of arteries and prone to further development (type IIa) and those at moderately susceptible sites that develop slowly, late, or not at all (type IIb). A lesion's position within the vascular tree and thus, the mechanical forces and the nature of the vessel wall at that point, determine, in large part, how a lesion is constituted." The intimal thickening at the lesion-prone position is called "adaptive thickening", and its existence is denied anywhere else in the arterial tree than the anatomically defined positions. The book you are now reading has over 200 pages of information about an entity that does not exist in conventional wisdom.

The counting of SMC's in the intima of human arteries has been reported only by the Moscow group, whose findings are reviewed by Chazov et. al. [9]. Cells were freed into suspension by an alcoholic-alkaline dissociation procedure. The total number of cells in the plaque were said to be increased 2.0-fold, after excluding foam cells. There is reason to think that their samples included an unspecified number of fibrous caps at atheromatous sites. They describe an envelope around SMC's. "The envelope usually contains several layers, each consisting of 130Å thick fibrils immersed in an amorphous substance of average electron density." This description invokes images of what is sometimes called "cocoon cells", by resemblance to a moth in a chrysalis. Such cells appear to be pathognomonic of atheroma whether they occur in artery, Achilles tendon, saphenous veins used in bypass grafts, or other structures; they seem to occur only when an atheronecrotic focus is nearby. Fibroplastic arterial sites always lack these cocoon cells in the absence of atheromatous necrotic core. Hence it seems possible that the increased SMC numbers reported by Chazov et. al. may reflect the proliferation of the fibrous cap that follows intrusion of atheroma. Those limited data offer little insight into the evolution of SMC numbers with aging leading up to atheroma.

Some work has been done on SMC numbers in experimental animals. Schwartz and Murry [43], for instance, emphasize the findings of Thomas et. al. [51]. "Thomas and coworkers claimed on the basis of elegant cell kinetic studies, that atherosclerotic lesions in fat-fed swine developed in preexisting intimal cell masses." This kind of information inclines us toward the view that an intima is needed to get intimal lesions in animals fed toxic levels of cholesterol, and a delay will be encountered if the normal intima is not already in place. The atheroma-prone cell masses in experimental animals are apt to be either normal intima acquired during development, or something analogous to it induced by wound healing. Evidence presented in this book indicates that humans have such "cell masses", i.e. normal intima, everywhere throughout the entire coronary macrovasculature by age 15 to 18 years, although it continues to mature at least until age 25 years. Again aging gets no attention in any such studies. This might be because the study of aging will require 30 or 40 or 50 years to mimic the human experience.

A consistently encountered finding in the data reviewed throughout this book is the early proliferation of SMC's as a developmental process during normal growth, and an absent or indolent proliferation when plaques begin to develop later in life. This formerly discredited idea is now gaining a degree of respectability. Schwartz & Murry [43] tell us that, "First, numerous studies of smooth muscle proliferation in human atherosclerotic lesions have failed to show evidence of extensive replication . . . If replication is not prominent at later stages of lesion development, then clonal expansion may be a very early event. The observation of low levels of replication in advanced plaques combined with evidence that the wall has the ability to develop large clonal patches even without atherosclerosis suggest that monoclonality could be the result of a proliferative burst in the intima over the first few months of life."

Schwartz & Murry address, in particular, the observation that the SMC's in atherosclerotic plaques (we may guess for the most part that they mean aggressive fibroplasia without atheroma) are chimeric with respect to heterozygous genes on the lyonized X chromosomes. Some plaques express chiefly one allele while other plaques express chiefly the other allele. The data in Chapter Five offer a possible explanation for this finding. Schwartz and Murry argue that the normal embryonic artery is chimeric, composed of a patchwork of clones from alternative choices for X chromosome inactivation. Therefore, during aggressive fibroplasia later in life, the enlargement of the realms of these monoclonal patches of SMC's greatly expands the volume occupied by the clone. If the thickness of a realm increases five fold, for instance, a commonplace finding, then the volume occupied by a clone might expand 10-20 fold. A sample of tissue limited to a minimum volume would far more often find monoclonality in the fibroplastic site than in the site of quiescent fibroplasia or normal intima.

Many of the concepts central to this treatise seem to be the same as those of Velican and Velican [77,78]. Those authors were among the last to study effects of aging upon human arteries before the attempt was generally abandoned. However, their failure to measure the sizes of the anatomic entities under investigation seriously hinders the understanding of what they try to say. Their unique terminology also aggravates this problem of communication. It is clear when they say, "The term 'atheroma' is in this report attributed to all necrotic plaques." That usage agrees with many authors including the author of this book. However, murkiness creeps in when they discuss "... non-raised and raised fibromuscular, mucoid, lipid-rich, and necrotic (atheroma) plaques." This leaves much uncertainty over what they mean by, "We are unable to reveal conversion of fatty streaks into atherosclerotic plaques." If we can take them to mean for fibromuscular plaque the kind of appearance here called aggressive fibroplasia, then they are describing a very commonplace entity that does not exist in modern conventional wisdom (Frame "3a" on this book cover). Furthermore, if by lipid-rich plaque they mean the collocating of fatty streak elements with aggressive fibroplasia, then our findings agree with theirs: the flat fatty streak (type IIb) does not evolve into atheroma by way of the AHA type III lesion (Figure 1-1D).

The existence of aggressive fibroplasia arising without participation by lipids is widely denied by modern literature on atherosclerosis. Yet this entity was generally accepted in the late 19th and early 20th centuries, not only as present but also as centrally important to the later entrapment of lipids to produce "fatty degenerations". It may be no coincidence that those early authors worked in Eastern Europe. Near the turn of the century, Ziegler, Thoma, Jores, Aschoff, and others looked at specimens in Berlin. Recently, Velican and Velican repeated many of the same observations in Bucharest. Perhaps the fibroplasia is more common there than in America and Western Europe, or perhaps, in that part of the world lipid is laggard in settling into the prepared sites. The latter possibility might seem plausible if blood lipid patterns are less atherogenic in that part of the world. The matter calls for investigation.

One aspect of the present treatise might seem to rouse little controversy, the sequestration of atherogenic lipids as a central event in atherogenesis. Williams and Tabas [81], for instance, note, "Although atherosclerosis is a complex and multifactorial process, we conclude that there exists a key pathogenic event, namely, lipoprotein retention, that is both necessary and sufficient to provoke lesion initiation in an otherwise-normal artery." However, the matter is not as clear as it may seem.

Those authors speak of "local vulnerability to lipoprotein retention" and "prelesional differences in lipoprotein retention between susceptible versus resistant arterial sites". Yet nowhere is any word to be found in their article on the distinguishing characteristics or the origins of such differences in vulnerability or susceptibility. The absence of these considerations abounds in recent literature. Authors may speak glibly of "prone" sites or "susceptible" sites, with no hint of how such proneness or susceptibility came to be there (other than by anatomic predestination), nor how its existence is to be diagnosed. Certainly no mention of a role for aging in these matters can ever be found. Nor have the "maleness" or "geography" risk factors ever been confronted as they are in the present treatise.

A large body of knowledge exists concerning arterial intimal matrix proteoglycans (PGs) and glycocosaminoglycans (GAGs) and their proclivity to aggregate with Apo-B100 bearing lipoproteins, forming aggregates that apparently also contain the enzyme lipoprotein lipase [7]. It is easy to envision that aggregations like these could account for the sequestration of lipids by SMC-depleted zones of aggressive fibroplasia. The little relevant information on this topic was recently reviewed [40], and this falls short of building a useful bridge between these two separate fields of inquiry. That is an obvious direction for future inquiry to consider.

The data reviewed in these pages are almost entirely those from personal experiences of the author. Readers who may be puzzled by the paucity of reference to data from other laboratories might note: The topics examined here concern a very thoroughly neglected area of pathology.

CHAPTER TWENTY SIX

REFERENCES

1. Allbutt C. (1911). Arteriosclerosis and the Kidney. *Brit Med J 1,* 853-858 & 922-927.
2. Anscombe F.J. (1950). Sampling theory of the negative binomial and logarithmic series distributions. *Biometrika 37,* 358-382.
3. Asai K., Kudej R.K., Shen Y.T.,Yang G.P., Takagi G., Kudej A.B., Geng Y.J., Sato N., Nazareno J.B., Vatner D.E., Natividad F., Bishop S.P., Vatner S.F. (2000). Peripheral vascular endothelial dysfunction and apoptosis in old monkeys. *Arterioscler ThrombVasc Biol 20,* 1493- 1499.
4. Bayless O.B, Adams C.W.M. (1980). The role of macrophages and giant cells in advanced human atherosclerosis. *Atherosclerosis 36,* 441-447.
5. Beck J.D., Elter J.R., Heiss G., Couper D., Mauriello S.M., Offenbacher S. (2001). Relationship of periodontal disease to carotid artery intima-media wall thickness. *Arterioscl Thromb Vasc Biol 21,* 1816-1822.
6. Berenson G.S., Srinivasan S.R., Bao W., Newman III W.P., Tracy R.E., Wattigney W.A. for the Bogalusa Heart Study. (1998). Association between multiple cardiovascular risk factors and atherosclerosis in children and young adults. *New Eng J Med 338,* 1650-1656.
7. Camejo G., Hurt-Camejo E., Wiklund O., Bondjers G. (1998) Association of apo B lipoproteins with arterial proteoglycans: Pathological significance and molecular basis. *Atherosclerosis 139, 205-222.*
8. Chait A. (1983).The role of lipoprotein receptors in lipid transport and in the pathogenesis of the hyperlipoproteinemias. *Special Topics Endocrin Metab 5,* 1-53.
9. Chazov E.I., Repin V.S., Orekhov A.N., Antonov A.S., Preobrazhensky S.N., Soboleva E.L., Smirnov V.N.. (1986). Atherosclerosis: What has been learned studying human arteries? *Atheroscl Rev 14,* 7-60.
10. Committee on vascular lesions of the council on arteriosclerosis, American Heart Association. (1992). A definition of the intima of human arteries and of its atherosclerosis- prone regions. *Arterioscl Thromb 12,* 120-134.
11. Committee on vascular lesions of the council on arteriosclerosis, American Heart Association. (1994). A definition of initial, fatty streak, and intermediate lesions of atherosclerosis. *Arterioscl Thromb 14,* 840-856.
12. Davies M.J., Richardson P.D., Woolf N., Katz D.R., Mann J. (1993). Risk of thrombosis in human atherosclerotic plaques: role of extracellular lipid, macrophage, and smooth muscle cell content. *Br Heart J 69,* 377-81.
13. DeStefano F., Anda R.F., Kahn H.S., Williamson D.F., Russell C.M. (1993). Dental disease and risk of coronary heart disease and mortality. *BMJ 306,* 688-691.
14. Dock W. (1946). The predilection of atherosclerosis for the coronary arteries. *JAMA 131, 875-878.*
15. Dohi Y., Kojima M., Sato K., & Luscher T.F. (1995). Age-related changes in vascular smooth muscle and endothelium. *Drugs & Aging 7,* 278-291.
16. Duff G.L. (1935). Experimental cholesterol arteriosclerosis and its relationship to

human arteriosclerosis. *Arch Pathol 20,* 81-123.

17. Duguid J.B. (1926). Atheroma of the aorta. J *Pathol Bact 29,* 371-387.
18. Falk E. (1983). Plaque rupture with severe pre-existing stenosis precipitating coronary thrombosis. *Br Heart J 50,* 127-34.
19. Fisher R.A. (1943). Species and the number of individuals in a random sample of an animal population. *J Animal Ecology 12,* 42-57.
20. Guyton J.R., & Klemp K.F. (1993). Transitional features in human atherosclerosis. *Amer J Pathol 143,* 1444-1457.
21. Higgins M., Kannel W., Garrison R., Pinsky J., Stokes III J. (1988). Hazards of obesity - the Framingham experience. *Acta Med Scand Suppl 723,* 23-36.
22. Hujoel P.P., Drangsholt M., Spiekerman C., DeRouen T.A. (2000). Periodontal disease and coronary heart disease risk. JAMA 284, 1406-1410.
23. Ikari Y., McManus B.M., Kenyon J., & Schwartz S.M. (1999). Neonatal intima formation in the human coronary artery. *Arterioscler Thromb Vasc Biol 19,* 2036-40.
24. Jones P.C.T., & Mollison J.E. (1948) A technique for the estimation of soil micro-organisms. *J General Microbiol 2,* 54-69.
25. Kiechl S., Willeit J. (for the Bruneck Study Group). (1999). The natural course of atherosclerosis, Part II: Vascular remodeling. *Arterioscl Thromb Vasc Biol 19,* 1491-1498.
26. Klotz O. (1910). Concerning compensatory hyperplasia of the intima. *J Exper Med 12,* 707-725.
27. Kortelainen M.L., & Sarkioja T. (1999). Extent and composition of coronary lesions in relation to fat distribution in women younger than 50 years of age. *Arterioscler Thromb Vasc Biol 19,* 695-699.
28. Lakka T.A., Lakka H.M., Salonen R., Kaplan G.A., Salonen J.T. (2001). Abdominal obesity is associated with accelerated progression of carotid atherosclerosis in men. *Atherosclerosis 154,* 497-504.
29. Manson J.E., Stampfer M.J., Hennekens C.H., Willett W.C. (1987). Body weight and longevity; A reassessment. *JAMA 257,* 353-358.
30. Masuda H., Zhuang Y.J., Singh T.M., Kawamura K., Murakami M., Zarins C.K., & Glagov S. (1999). Adaptive remodeling of internal elastic lamina and endothelial lining during flow- induced arterial enlargement. *Arterioscler Thromb Vasc Biol 19,* 2298-307.
31. Mattila K.J., Nieminen M.S., Valtonen V.V., Rasi V.P., Kesaniemi Y.A., Syrjala S.L., Jimgell P.S., Isoluoma M., Hietaniemi K., Jokinen M.J., Huttunen J.K. (1989). Association between dental health and acute myocardial infarction. *BMJ 298,* 779-781.
32. Mattila K.J., Asikainen S., Wolf J., Jousimies-Somer H., Valtonen V., Nieminen M. (2000). Age, Dental infections, and coronary heart disease. *J Dent res 79,* 756-760.
33. McFarlane M.J., Feinstein A.R., Wells C.K., Chan C.K. (1987). The 'epidemiologic necropsy'. *JAMA 258,* 331-338.
34. McFarlane M.J. (1991). The epidemiologic necropsy for abdominal aneurysm. *JAMA 265,* 2085-2088.

35. McGill H.C., McMahan C.A., Malcom G.T., Oalmann M.C., Strong J.P., and the Pathobiological Determinants of Atherosclerosis in Youth (PDAY) Research Group. (2002). Obesity accelerates the progression of atherosclerosis in young men. *Circ 105,* 2712-2718.
36. McGill HC, McMahan CA, Malcom GT, Oalmann MC, Strong JP, and the Pathobiological Determinants of Atherosclerosis in Youth (PDAY) Research Group. (1995). Relation of glycohemoglobin and adiposity to atherosclerosis in youth. *Arterioscler Thromb Vasc Biol 15,* 431-440.
37. McMahan C.A. (1968). Autopsied cases by age, sex, and "Race". *Lab Invest 18,* 468-478.
38. Plenz G., Dorszewski A., Breithard G., Robenek H. (1999). Expression of Type VIII collagen after cholesterol diet and injury in the rabbit model of atherosclerosis. *Arterioscler Thromb Vasc Biol 19,* 1201-1209.
39. Quenouille M.H. (1949). A relation between the logarithmic, Poisson, and negative binomial series. *Biometrics 5,* 162-164.
40. Radhakrishnamurthy B., Tracy R.E., Delfares E.R. Jr., Berenson G.S. (1998) Proteoglycans in human coronary arteriosclerotic lesions. Exp Molec Path 65, 1-8.
41. Rekhter M.D., Andreeva E.R., Mironov A.A., Orekhov A.N. (1991). Three-dimensional cytoarchitecture of normal and atherosclerotic intima of human aorta. *Amer J Pathol 138,* 569-580.
42. Restrepo C., Eggen D.A., Guzman M.A., Tejada C. (1973). Postmortem dimensionsof the coronary arteries in different geographic locations. *Lab Invest 28,* 244-51.
43. Schwartz S.M., & Murry C.E. (1998). Proliferation and the monoclonal origins of atherosclerotic lesions. *Ann Rev Med 49,* 437-460.
44. Seymour R.A. (2001). Heart of the matter. *Brit Dent J 190,* 106.
45. Spring P.M., & Hoff H.F. (1989). LDL accumulation in the grossly normal human iliac bifurcation and common iliac arteries. *Exper Molec Pathol 51,* 179-185.
46. Solberg L.A. & Strong J.P. (1983). Risk factors and atherosclerotic lesions. *Arterioscler 3,* 187-198.
47. Stary H.C. (1984). Location and development of atherosclerotic lesions in coronary arteries. *Atherosclerosis 50,* 237-239.
48. Stary H.C. (1992). Composition and classification of human atherosclerotic lesions. *Virchows Arch 421,* 277-290.
49. Stary H.C. (2000). Natural history and histological classification of atherosclerotic lesions an update. *Arterioscler Thromb Vasc Biol 20,* 1177-1178.
50. Temelkova-Kurktschiev T., Koehler C., Schaper F., Henkel E., Hahnefeld A., Fuecker K., Siegert G., Hanefeld M. (1998). Relationship between fasting plasma glucose, atherosclerosis risk factors and carotid intima media thickness in non-diabetic individuals. *Diabetologia 41,* 706-712.
51. Thomas W.A., Reiner J.M., Florentin R.A., Scott R.F. (1979). Population dynamics of arterial cells during atherogenesis. VIII. Separation of the roles of injury and growth stimulation in early aortic atherogenesis in swine originating in pre-existing intimal smooth muscle cell masses. *Exp Mol Pathol 31,* 124-144.

52. Tracy, R.E. (1995). Declining density of intimal smooth muscle cells and age as preconditions for atheronecrosis in the basilar artery. *Virchows Arch 427,* 131-138.
53. Tracy R.E. (1997). Declining density of intimal smooth muscle cells as a precondition for atheronecrosis in the coronary artery. *Virchows Arch 430,* 155-162.
54. Tracy, R.E. (1998). Histological characteristics of coronary artery in relation to the renovasculopathies of hypertension. *Ann Diag Pathol 2,* 1-9.
55. Tracy R.E. (1998). Medial thickness of coronary arteries as a correlate of atherosclerosis. *Atherosclerosis 139,* 11-19.
56. Tracy R.E. (2001). Medial thickenings of coronary artery and the aging risk factor for atherosclerosis. *Atherosclerosis 155,* 337-346.
57. Tracy R.E., Berenson G.S., Cueto-Garcia L., Wattigney W.A., Barrett T.J. (1992). Nephrosclerosis and aortic atherosclerosis from age 6 to 70 years in the United States and Mexico. *Virch Arch A Pathol Anat 420,* 479-488.
58. Tracy R.E. & Ishii T. (2000). What is "nephrosclerosis"? Lessons from the USA Japan, and Mexico. *Nephrol Dial Transpl 15,* 1357-1366.
59. Tracy, R.E., & Johnson, L.K. (1994). Aging of a class of arteries in various mammalian species in relation to life span. *Gerontology 40,* 291-297.
60. Tracy, R.E., & Kissling, G.E. (1985). Age and fibroplasia as preconditions for atheronecrosis in the human thoracic aorta. *Arch Path Lab Med 109,* 651-658.
61. Tracy, R.E., & Kissling, G.E. (1987). Age and fibroplasia as preconditions for atheronecrosis in human coronary arteries. *Arch Path Lab Med 111,* 957-963.
62. Tracy R.E., Kissling G.E., & Curtis M.B. (1987). Smooth muscle cell-reticulin lamellar units of 13.2 μm thickness composing the aortic intima. *Virchows Arch 411,* 415-424.
63. Tracy R.E., Kissling G.E., Devaney K.O., Lopez C.R., Gandia M. (1985). Spatial dispersion of spindle cells in the human thoracic aorta. *Lab Invest 52,* 85-92.
64. Tracy R.E., Kissling G.E., Gandia M., Reynolds C. (1989). Spatial dispersion of stainable lipid in frozen sections of human aorta. *Virchows Archiv 414,* 39-49.
65. Tracy, R.E., Kissling, G.E., Lopez, C.R., Gandia, M. (1986). Histometric components of aortic atherosclerosis which vary or remain constant among eight populations. *Lab Invest 54,* 314-321.
66. Tracy R.E., Kissling G.E., Malcom G.T., & Devaney K.O. (1987). Sequestration hypothesis of atherosclerosis. *Virchow's Arch 411,* 425-434.
67. Tracy, R.E., Kissling, G.E., Oalmann, M.C. (1989). Two variants of coronary atherosclerosis in six populations. *Amer J Publ Health 79,* 575-581.
68. Tracy, R.E., Malcom, G.T., Oalmann, M.C., Qureshi, U., Ishii, T., Velez-Duran, M. (1992). Renal microvascular features of hypertension, in Japan, Guatemala, and USA. *Arch Pathol Lab Med 116,* 50-55.
69. Tracy, R.E., Mercante D.E., Moncada A., Berenson G. (1986). Quantitation of hypertensive nephrosclerosis on an objective scale of measure in adults and children. *Amer J Clin Pathol 85,* 312-318.
70. Tracy R.E., Mulvad G., Pederson H.S., Jul, E., Bjerregaard, P., Newmann, W.P. III. (1996). Blood pressure in people in Greenland assessed by measuring renovasculopathies of hypertension at autopsy. *Amer J Hypertens 9,* 560-565.

71. Tracy, R.E., Parra, D., Eisaguirre, W., Torres-Balanza, R.A. (2002). Influence of arteriolar hyalinization on arterial intimal fibroplasia in the renal cortex of subjects in the USA, Peru, and Bolivia, applicable also to other populations. *Amer J Hypertens In Press.*
72. Tracy, R.E., Strong, J.P., Toca, V.T., Lopez, C.R. (1979). Atheronecrosis and its fibroproliferative base and cap in the thoracic aorta. *Lab Invest 41,* 546-552.
73. Tracy, R.E., Strong, J.P., Toca, V.T., Lopez, C.R. (1979). Variable patterns of nonatheromatous aortic intimal thickening. *Lab Invest 41,* 553-559.
74. Tracy, R.E., Toca, V.T., Kissling, G.E., & Lopez, C. 1983. Fibrous intimal thickening and atheronecrosis of the thoracic aorta; Some background for mathematical theory. Lab Invest 48, 313-322.
75. Tracy, R.E. & Walia, P. 2002. A method to fix lipids for staining fat embolism in paraffin sections. *Histopath 40,* 1-5.
76. Tracy, R.E., & White S. (2002). A method for quantifying adrenocortical nodular hyperplasia at autopsy: Some use of the method in illuminating hypertension and atherosclerosis. *Ann Diag Pathol 6,* 20-29.
77. Velican, C., & Velican D. (1976). Intimal thickening in developing coronary arteries and its relevance to atherosclerotic involvement. Atherosclerosis 23, 345-355.
78. Velican C., & Velican D. (1980). The precursors of coronary atherosclerotic plaques in subjects up to 40 years old. *Atherosclerosis 37,* 33-46.
79. Virmani R., Kolodgie F.D., Burke A.P., Farb A., Schwartz S.M. (2000). Lessons from sudden coronary death: A comprehensive morphological classification scheme for atherosclerotic lesions. *Arterioscler Thromb Vasc Biol 20,* 1262-1275.
80. Wilens S.L. (1951). The nature of diffuse intimal thickening of arteries. *Amer J Pathol 27,* 825-839.
81. Williams K.J., & Taas I. (1995). The response-to-retention hypothesis of early atherogenesis. *Arterioscler Thromb Vasc Biol 15,* 551-561.

MIX
Papier aus verantwortungsvollen Quellen
Paper from responsible sources
FSC® C105338

If you have any concerns about our products,
you can contact us on
ProductSafety@springernature.com

In case Publisher is established outside the EU,
the EU authorized representative is:
Springer Nature Customer Service Center GmbH
Europaplatz 3, 69115 Heidelberg, Germany

Printed by Libri Plureos GmbH
in Hamburg, Germany